Answer Key

SRA Corrective Mathematics

Subtraction

A Direct Instruction Program

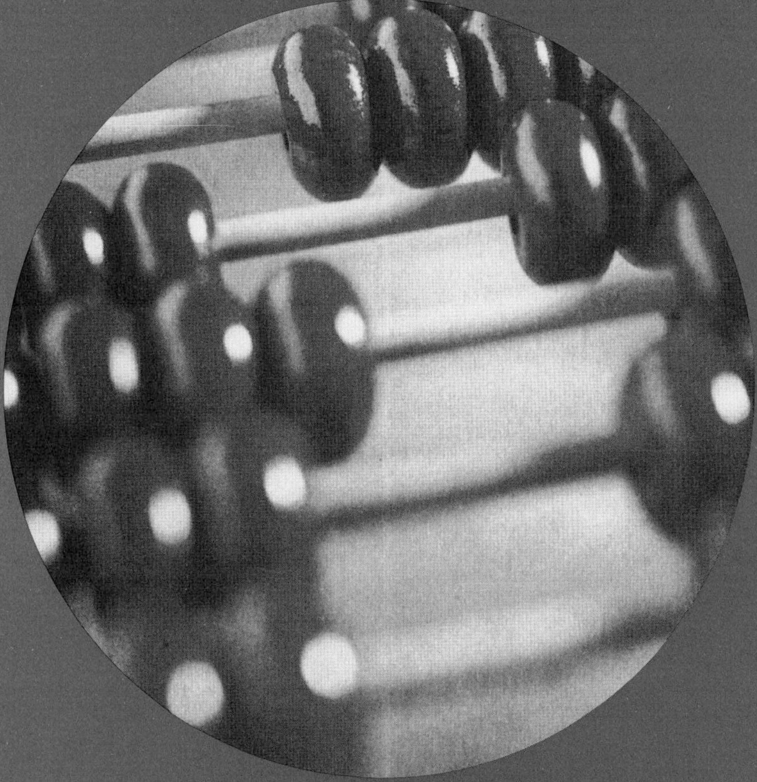

Siegfried Engelmann • Doug Carnine

SRA
Columbus, OH

The McGraw·Hill Companies

Cover and title page photo credits:
©Steve Mercer/Getty Images, Inc.

SRAonline.com

SRA

Send all inquiries to:
SRA/McGraw-Hill
4400 Easton Commons
Columbus, OH 43219

Printed in the United States of America.

ISBN 0-07-602463-6

5 6 7 8 9 10 11 GLO 15 14 13 12 11 10 09

The McGraw·Hill Companies

Answer Key
Subtraction Contents

Subtraction Preskill Test

A

8	7	8	7	6	5	8
+ 9	+ 5	+ 4	+ 9	+ 8	+ 9	+ 3
17	12	12	16	14	14	11
7	6	8	9	8	5	7
+ 6	+ 6	+ 7	+ 2	+ 8	+ 8	+ 4
13	12	15	11	16	13	11
8	9	7	9	6	9	8
+ 5	+ 4	+ 7	+ 7	+ 5	+ 9	+ 4
13	13	14	16	11	18	12

B

351	47	605	86	264
+ 120	+ 25	+ 123	+ 13	+ 183
471	72	728	99	447

C

Ann buys 5 tires. She buys 2 more tires. How many tires in all did Ann buy?

5 + 2 = 7

George found 6 cans. He found 3 more cans. How many cans in all did George find?

6 + 3 = 9

Chris got 4 pens. She got 1 more pen. How many pens in all did Chris get?

4 + 1 = 5

Jose made 7 chairs. He made 2 more chairs. How many chairs in all did Jose make?

7 + 2 = 9

Subtraction Preskill Test

A

8	7	8	7	6	5	8
+ 9	+ 5	+ 4	+ 9	+ 8	+ 9	+ 3
17	12	12	16	14	14	11
7	6	8	9	8	5	7
+ 6	+ 6	+ 7	+ 2	+ 8	+ 8	+ 4
13	12	15	11	16	13	11
8	9	7	9	6	9	8
+ 5	+ 4	+ 7	+ 7	+ 5	+ 9	+ 4
13	13	14	16	11	18	12

B

351	47	605	86	264
+ 120	+ 25	+ 123	+ 13	+ 183
471	72	728	99	447

C

Ann buys 5 tires. She buys 2 more tires. How many tires in all did Ann buy?

5 + 2 = 7

George found 6 cans. He found 3 more cans. How many cans in all did George find?

6 + 3 = 9

Chris got 4 pens. She got 1 more pen. How many pens in all did Chris get?

4 + 1 = 5

Jose made 7 chairs. He made 2 more chairs. How many chairs in all did Jose make?

7 + 2 = 9

Same as page a

Subtraction Placement Test

A

7	8	5	6	6
− 1	− 4	− 1	− 0	− 3
6	4	4	6	3
10	4	9	4	6
− 5	− 0	− 1	− 2	− 1
5	4	8	2	5

B

7 1	4 1	4 1	1 1	5 1	2 1
82	54	526	426	62	347
− 35	− 37	− 165	− 318	− 5	− 62
47	17	361	108	57	285

C

There are 148 red cars and 432 blue cars. How many more blue cars are there than red cars?

3 1 2 1
432
− 148
284 cars

The shop gave away 86 red balloons. The shop gave away 90 blue balloons. How many balloons in all did the shop give away?

86
+ 90
176 balloons

Part C continues on the next page.

Subtraction Placement Test (continued)

Lucy found 164 pencils. 98 of the pencils were broken. How many of the pencils were not broken?

```
 0 15 1
  1̶8̶4
 − 98
 ----
  66 pencils
```

1840 girls go to our school. There are 3000 children altogether in our school. How many boys go to our school?

```
  2 9 1
  3̶0̶0̶0
− 1840
------
  1160 boys
```

When Bill started high school, he weighed 108 kilograms. He has gained 24 kilograms since then. How many kilograms does Bill weigh now?

```
  108
+  24
-----
  132 kilograms
```

At the beginning of the week, Jenny had 945 points. At the end of the week, Jenny had 1000 points. How many points did Jenny get during the week?

```
   9 9 1
  1̶0̶0̶0
−  945
------
    55 points
```

The truck is carrying 1463 kilograms of apples and 3652 kilograms of oranges. How many kilograms of fruit is the truck carrying?

```
  1463
+ 3652
------
  5115 kilograms
```

The old car weighs 986 kilograms. The new car weighs 2000 kilograms. How much heavier is the new car?

```
 1 9 9 1
  2̶0̶0̶0
−  986
------
  1014 kilograms
```


Lesson 1

Bonus

1

| | | | |

2

A	B	C	D	E	F
264	538	492	751	892	376

3

A. [7] { 4: 7 − 4 = 3; 3: 7 − 3 = 4 }

B. [9] { 2: 9 − 2 = 7; 7: 9 − 7 = 2 }

C. [3] { 2: 3 − 2 = 1; 1: 3 − 1 = 2 }

D. [5] { 3: 5 − 3 = 2; 2: 5 − 2 = 3 }

Lesson 2

Bonus

1

A	B	C	D	E
279	146	657	328	864

2

| | | | | |

3

4	6	9	5	8	10	7	2
− 1	− 1	− 1	− 1	− 1	− 1	− 1	− 1
3	5	8	4	7	9	6	1

4

A. [8] { 5: 8 − 5 = 3; 3: 8 − 3 = 5 }

B. [6] { 4: 6 − 4 = 2; 2: 6 − 2 = 4 }

C. [5] { 3: 5 − 3 = 2; 2: 5 − 2 = 3 }

D. [7] { 2: 7 − 2 = 5; 5: 7 − 5 = 2 }

5

A. 5 − 5 = 0

B. 5 − 4 = 1

C. 5 − 3 = 2

D. 5 − 2 = 3

E. 5 − 1 = 4

Lesson 3

Bonus

1

A	B	C	D	E
425	738	916	437	514

2

5	7	10	8	4	9	6	2
− 1	− 1	− 1	− 1	− 1	− 1	− 1	− 1
4	6	9	7	3	8	5	1

3

A. [5] { 4 + 1 = 5; 1: 5 − 1 = 4; 4: 5 − 4 = 1 }

B. [7] { 6 + 1 = 7; 1: 7 − 1 = 6; 6: 7 − 6 = 1 }

C. [9] { 8 + 1 = 9; 1: 9 − 1 = 8; 8: 9 − 8 = 1 }

D. [4] { 3 + 1 = 4; 1: 4 − 1 = 3; 3: 4 − 3 = 1 }

4

A. 6 − 6 = 0

B. 6 − 5 = 1

C. 6 − 4 = 2

D. 6 − 3 = 3

E. 6 − 2 = 4

F. 6 − 1 = 5

Lesson 4

Bonus

1

A	B	C	D	E	F
356	297	424	932	358	672

2

A. 10 { 5, 5 }
- 5 + 5 = 10
- 10 − 5 = 5

B. 18 { 9, 9 }
- 9 + 9 = 18
- 18 − 9 = 9

C. 6 { 3, 3 }
- 3 + 3 = 6
- 6 − 3 = 3

D. 8 { 4, 4 }
- 4 + 4 = 8
- 8 − 4 = 4

3

6	8	2	9	4	7	10	3
− 1	− 1	− 1	− 1	− 1	− 1	− 1	− 1
5	7	1	8	3	6	9	2

4

A. 5 { 1, 4 }
- 4 + 1 = 5
- 5 − 1 = 4
- 5 − 4 = 1

B. 8 { 1, 7 }
- 7 + 1 = 8
- 8 − 1 = 7
- 8 − 7 = 1

C. 10 { 1, 9 }
- 9 + 1 = 10
- 10 − 1 = 9
- 10 − 9 = 1

D. 4 { 1, 3 }
- 3 + 1 = 4
- 4 − 1 = 3
- 4 − 3 = 1

Lesson 5

Facts + Bonus = TOTAL

1

A	B	C	D	E	F
406	308	728	507	346	205

2

A. 4 { 2, 2 }
- 2 + 2 = 4
- 4 − 2 = 2

B. 10 { 5, 5 }
- 5 + 5 = 10
- 10 − 5 = 5

C. 18 { 9, 9 }
- 9 + 9 = 18
- 18 − 9 = 9

D. 6 { 3, 3 }
- 3 + 3 = 6
- 6 − 3 = 3

3

10	6	6	18	6	8	7	18
− 5	− 3	− 1	− 9	− 3	− 1	− 1	− 9
5	3	5	9	3	7	6	9
2	10	6	8	18	10	10	7
− 1	− 5	− 3	− 1	− 9	− 5	− 1	− 1
1	5	3	7	9	5	9	6

4

A. 7 { 1, 6 }
- 6 + 1 = 7
- 7 − 1 = 6
- 7 − 6 = 1

B. 10 { 1, 9 }
- 9 + 1 = 10
- 10 − 1 = 9
- 10 − 9 = 1

C. 9 { 1, 8 }
- 8 + 1 = 9
- 9 − 1 = 8
- 9 − 8 = 1

D. 4 { 1, 3 }
- 3 + 1 = 4
- 4 − 1 = 3
- 4 − 3 = 1

Lesson 6

Facts + Bonus = TOTAL

1

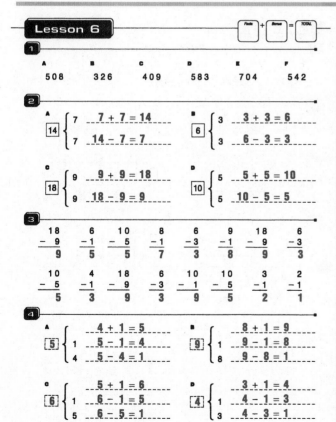

A	B	C	D	E	F
508	326	409	583	704	542

2

A. 14 { 7, 7 }
- 7 + 7 = 14
- 14 − 7 = 7

B. 6 { 3, 3 }
- 3 + 3 = 6
- 6 − 3 = 3

C. 18 { 9, 9 }
- 9 + 9 = 18
- 18 − 9 = 9

D. 10 { 5, 5 }
- 5 + 5 = 10
- 10 − 5 = 5

3

18	6	10	8	6	9	18	6
− 9	− 1	− 5	− 1	− 3	− 1	− 9	− 3
9	5	5	7	3	8	9	3
10	4	18	6	10	10	3	2
− 5	− 1	− 9	− 3	− 1	− 5	− 1	− 1
5	3	9	3	9	5	2	1

4

A. 5 { 1, 4 }
- 4 + 1 = 5
- 5 − 1 = 4
- 5 − 4 = 1

B. 9 { 1, 8 }
- 8 + 1 = 9
- 9 − 1 = 8
- 9 − 8 = 1

C. 6 { 1, 5 }
- 5 + 1 = 6
- 6 − 1 = 5
- 6 − 5 = 1

D. 4 { 1, 3 }
- 3 + 1 = 4
- 4 − 1 = 3
- 4 − 3 = 1

Lesson 7

Facts + Bonus = TOTAL

1

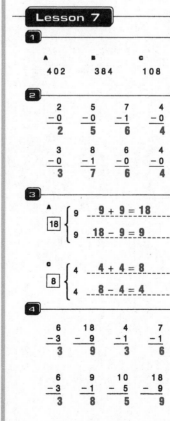

A	B	C	D	E
402	384	108	530	714

2

2	5	7	4	8	6	8	4
− 0	− 0	− 1	− 0	− 1	− 0	− 0	− 1
2	5	6	4	7	6	8	3
3	8	6	4	10	3	6	2
− 0	− 1	− 0	− 0	− 1	− 1	− 0	− 0
3	7	6	4	9	2	6	2

3

A. 18 { 9, 9 }
- 9 + 9 = 18
- 18 − 9 = 9

B. 14 { 7, 7 }
- 7 + 7 = 14
- 14 − 7 = 7

C. 8 { 4, 4 }
- 4 + 4 = 8
- 8 − 4 = 4

D. 12 { 6, 6 }
- 6 + 6 = 12
- 12 − 6 = 6

4

6	18	4	7	10	10	18	6
− 3	− 9	− 1	− 1	− 1	− 5	− 9	− 1
3	9	3	6	9	5	9	5
6	9	10	18	6	10	10	9
− 3	− 1	− 5	− 9	− 3	− 1	− 5	− 1
3	8	5	9	3	9	5	8

Lesson 7 (continued)

5

A. [9] { 1: 9 − 1 = 8; 8: 9 − 8 = 1 }

B. [7] { 1: 7 − 1 = 6; 6: 7 − 6 = 1 }

C. [10] { 1: 10 − 1 = 9; 9: 10 − 9 = 1 }

D. [4] { 1: 4 − 1 = 3; 3: 4 − 3 = 1 }

E. [6] { 1: 6 − 1 = 5; 5: 6 − 5 = 1 }

F. [8] { 1: 8 − 1 = 7; 7: 8 − 7 = 1 }

D
1324

Lesson 8

Facts + Bonus = TOTAL

1

15	13	14	6	15	11	13	12
− 10	− 10	− 10	− 0	− 10	− 10	− 10	− 10
5	3	4	2	5	1	3	2

2

A	B	C	D	E
3456	4546	2538	1324	8260

3

6	4	3	8	10	6	4	9
− 0	− 0	− 1	− 0	− 1	− 0	− 1	− 0
6	4	2	8	9	6	3	9

4

A. [8] { 4: 4 + 4 = 8; 4: 8 − 4 = 4 }

B. [18] { 9: 9 + 9 = 18; 9: 18 − 9 = 9 }

C. [14] { 7: 7 + 7 = 14; 7: 14 − 7 = 7 }

D. [10] { 5: 5 + 5 = 10; 5: 10 − 5 = 5 }

5

14	8	18	10	6	14	18	8
− 7	− 4	− 9	− 5	− 3	− 7	− 9	− 4
7	4	9	5	3	7	9	4

Lesson 8 (continued)

6

14	8	6	10	18	8	6	14
− 7	− 4	− 3	− 5	− 9	− 1	− 1	− 7
7	4	3	5	9	7	5	7
3	18	8	8	6	4	14	6
− 1	− 9	− 1	− 4	− 3	− 1	− 7	− 1
2	9	7	4	3	3	7	5
10	8	14	6	18	7	10	10
− 5	− 4	− 7	− 3	− 9	− 1	− 1	− 5
5	4	7	3	9	6	9	5
8	7	10	6	14	8	6	8
− 1	− 1	− 5	− 1	− 7	− 4	− 3	− 1
7	6	5	5	7	4	3	7

7

A. [8] { 0: 8 − 0 = 8; 8: 8 − 8 = 0 }

B. [6] { 0: 6 − 0 = 6; 6: 6 − 6 = 0 }

C. [4] { 1: 4 − 1 = 3; 3: 4 − 3 = 1 }

D. [9] { 0: 9 − 0 = 9; 9: 9 − 9 = 0 }

E. [6] { 1: 6 − 1 = 5; 5: 6 − 5 = 1 }

F. [6] { 0: 6 − 0 = 6; 6: 6 − 6 = 0 }

Lesson 9

Facts + Bonus = TOTAL

1

A	B	C	D	E
4285	3624	8153	9284	7295

2

15	12	14	11	13	15	12	11
− 10	− 10	− 10	− 10	− 10	− 10	− 10	− 10
5	2	4	1	3	5	2	1

3

8	4	7	3	10	5	8	10
− 0	− 1	− 0	− 0	− 1	− 0	− 1	− 0
8	3	7	3	9	5	7	10

4

A. [14] { 7: 7 + 7 = 14; 7: 14 − 7 = 7 }

B. [16] { 8: 8 + 8 = 16; 8: 16 − 8 = 8 }

C. [18] { 9: 9 + 9 = 18; 9: 18 − 9 = 9 }

D. [12] { 6: 6 + 6 = 12; 6: 12 − 6 = 6 }

5

14	10	18	8	7	10	6	6
− 7	− 1	− 9	− 4	− 0	− 5	− 0	− 3
7	9	9	4	7	5	6	3
10	18	10	8	8	14	6	5
− 0	− 9	− 5	− 4	− 0	− 7	− 3	− 1
10	9	5	4	8	7	3	4

Part 5 continues on the next page.

Lesson 9 (continued)

8	6	14	3	18	8	6	6
− 4	− 1	− 7	− 0	− 9	− 1	− 3	− 0
4	5	7	3	9	7	3	6
9	9	6	6	6	8	5	8
− 1	− 0	− 3	− 1	− 0	− 4	− 1	− 0
8	9	3	5	6	4	4	8

6

A [3] { 1: 3 − 1 = 2; 2: 3 − 2 = 1 }

B [7] { 0: 7 − 0 = 7; 7: 7 − 7 = 0 }

C [4] { 0: 4 − 0 = 4; 4: 4 − 4 = 0 }

D [9] { 1: 9 − 1 = 8; 8: 9 − 8 = 1 }

E [9] { 0: 9 − 0 = 9; 9: 9 − 9 = 0 }

F [7] { 1: 7 − 1 = 6; 6: 7 − 6 = 1 }

G [5] { 1: 5 − 1 = 4; 4: 5 − 4 = 1 }

H [8] { 0: 8 − 0 = 8; 8: 8 − 8 = 0 }

Lesson 10

Facts + Bonus = TOTAL

1

A 3279 B 146 C 6571 D 432 E 2384

2

14	17	13	15	18	12	16	19
− 10	− 10	− 10	− 10	− 10	− 10	− 10	− 10
4	7	3	5	8	2	6	9

3

7	9	4	8	3	6	10	10
− 0	− 0	− 1	− 0	− 1	− 0	− 1	− 0
7	9	3	8	2	6	9	10

4

A [18] { 9: 9 + 9 = 18; 9: 18 − 9 = 9 }

B [8] { 4: 4 + 4 = 8; 4: 8 − 4 = 4 }

C [10] { 5: 5 + 5 = 10; 5: 10 − 5 = 5 }

D [6] { 3: 3 + 3 = 6; 3: 6 − 3 = 3 }

E [12] { 6: 6 + 6 = 12; 6: 12 − 6 = 6 }

F [14] { 7: 7 + 7 = 14; 7: 14 − 7 = 7 }

5

14	10	18	8	6	14	10	18
− 7	− 5	− 9	− 4	− 3	− 7	− 5	− 9
7	5	9	4	3	7	5	9

Lesson 10 (continued)

6

18	14	10	8	6	8	18	14
− 9	− 7	− 5	− 0	− 3	− 4	− 9	− 7
9	7	5	8	3	4	9	7
7	5	14	10	10	10	8	8
− 0	− 1	− 7	− 1	− 0	− 5	− 1	− 4
7	4	7	9	10	5	7	4
8	18	8	10	14	5	7	3
− 0	− 9	− 4	− 5	− 7	− 0	− 0	− 1
8	9	4	5	7	5	7	2
5	8	7	9	4	18	14	8
− 1	− 4	− 0	− 1	− 2	− 9	− 7	− 4
4	4	7	8	2	9	7	4

7

A [7] { 0: 7 − 0 = 7; 7: 7 − 7 = 0 }

B [4] { 1: 4 − 1 = 3; 3: 4 − 3 = 1 }

C [5] { 0: 5 − 0 = 5; 5: 5 − 5 = 0 }

D [9] { 1: 9 − 1 = 8; 8: 9 − 8 = 1 }

E [4] { 0: 4 − 0 = 4; 4: 4 − 4 = 0 }

F [8] { 1: 8 − 1 = 7; 7: 8 − 7 = 1 }

Lesson 11

Facts + Problems + Bonus = TOTAL

1

A 3856 B 4258 C 326 D 1420 E 350 F 4263

2

14	18	15	19	13	17	12	16
− 10	− 10	− 10	− 10	− 10	− 10	− 10	− 10
4	8	5	9	3	7	2	6

3

A [18] { 9: 9 + 9 = 18; 9: 18 − 9 = 9 }

B [14] { 7: 7 + 7 = 14; 7: 14 − 7 = 7 }

C [16] { 8: 8 + 8 = 16; 8: 16 − 8 = 8 }

D [10] { 5: 5 + 5 = 10; 5: 10 − 5 = 5 }

4

18	14	10	10	8	8	14	18
− 9	− 7	− 5	− 1	− 0	− 4	− 7	− 9
9	7	5	9	8	4	7	9
9	9	6	5	10	6	18	7
− 0	− 1	− 3	− 1	− 5	− 1	− 9	− 0
9	8	3	4	5	5	9	7
18	7	8	3	18	8	8	14
− 9	− 0	− 4	− 0	− 9	− 1	− 4	− 7
9	7	4	3	9	7	4	7

Lesson 11 (continued)

5

A. [7] { 1: 7 − 1 = 6; 6: 7 − 6 = 1 }

B. [5] { 0: 5 − 0 = 5; 5: 5 − 5 = 0 }

C. [4] { 0: 4 − 0 = 4; 4: 4 − 4 = 0 }

D. [10] { 1: 10 − 1 = 9; 9: 10 − 9 = 1 }

6

10	9	7	4	8	5	9	10
−10	−8	−6	−4	−7	−5	−9	− 9
0	**1**	**1**	**0**	**1**	**0**	**0**	**1**

5	3	8	3	5	7	2	6
−4	−3	−8	−2	−4	−7	−2	−5
1	**0**	**0**	**1**	**1**	**0**	**0**	**1**

7

A. $\overset{3}{\cancel{4}}{}^{1}59$ B. $\overset{6}{\cancel{7}}{}^{1}4$ C. $\overset{2}{\cancel{3}}{}^{1}572$ D. $\overset{7}{\cancel{8}}{}^{1}40$

8

A	B	C
841	537	328
−410	−101	−104
431	**436**	**224**

9

A	B	C	D	E
546	546	4863	39	5263
−101	+101	−2100	−11	+1102
445	**647**	**2763**	**28**	**6365**

Lesson 12

Test + Facts + Problems + Bonus = TOTAL

1

15	18	13	17	15	12	14	19
−10	−10	−10	−10	−10	−10	−10	−10
5	**8**	**3**	**7**	**5**	**2**	**4**	**9**

2

A. 4236 B. 8150 C. 425 D. 7111 E. 711 F. 250

3

A. [5] { 1: 5 − 1 = 4; 4: 5 − 4 = 1 }

B. [5] { 0: 5 − 0 = 5; 5: 5 − 5 = 0 }

C. [8] { 0: 8 − 0 = 8; 8: 8 − 8 = 0 }

D. [10] { 1: 10 − 1 = 9; 9: 10 − 9 = 1 }

E. [8] { 1: 8 − 1 = 7; 7: 8 − 7 = 1 }

F. [4] { 0: 4 − 0 = 4; 4: 4 − 4 = 0 }

4

10	4	14	8	14	2	9	5
−10	−3	− 6	−7	− 8	−2	−8	−4
0	**1**	**8**	**1**	**6**	**0**	**1**	**1**

10	5	3	14	4	14	3	6
− 9	−5	−2	− 8	−4	− 6	−2	−5
1	**0**	**1**	**6**	**0**	**8**	**1**	**1**

Lesson 12 (continued)

5

8	4	7	8	3	9	5	8
−0	−0	−1	−0	−1	−0	−0	−1
8	**4**	**6**	**8**	**2**	**9**	**5**	**7**

6

A. $6\overset{3}{\cancel{4}}{}^{1}8$ B. $37\overset{8}{\cancel{9}}{}^{1}0$ C. $\overset{6}{\cancel{7}}{}^{1}354$ D. $\overset{0}{\cancel{1}}{}^{1}48$ E. $5\overset{8}{\cancel{9}}{}^{1}35$

7

A	B	C
496	638	345
−200	−104	−121
296	**534**	**224**

8

A	B	C	D
985	985	452	488
−101	+901	+201	−101
884	**1886**	**653**	**387**

E	F	G	H
472	759	5263	485
−200	+110	+1102	−120
272	**869**	**6365**	**365**

Lesson 13

Facts + Problems + Bonus = TOTAL

1

A. 4000 B. 400 C. 318 D. 1150 E. 2314 F. 107

2

14	14	14	14	14	14	14	14
− 8	−10	− 6	− 7	− 8	−10	− 6	− 7
6	**4**	**8**	**7**	**6**	**4**	**8**	**7**

3

12	12	12	12	12	12	12	12
− 3	− 4	− 5	− 4	− 3	− 5	− 4	− 5
9	**8**	**7**	**8**	**9**	**7**	**8**	**7**

4

A. [8] { 7: 8 − 7 = 1; 1: 8 − 1 = 7 }

B. [9] { 9: 9 − 9 = 0; 0: 9 − 0 = 9 }

C. [4] { 4: 4 − 4 = 0; 0: 4 − 0 = 4 }

D. [6] { 5: 6 − 5 = 1; 1: 6 − 1 = 5 }

5

8	6	4	7	3	9	10	8
−8	−5	−3	−7	−3	−8	− 9	−8
0	**1**	**1**	**0**	**0**	**1**	**1**	**0**

9	7	5	8	3	6	10	10
−9	−6	−5	−7	−2	−6	−10	− 9
0	**1**	**0**	**1**	**1**	**0**	**0**	**1**

Lesson 13 (continued)

6

A 12 { 6, 6 — 6 + 6 = 12; 12 − 6 = 6

B 14 { 7, 7 — 7 + 7 = 14; 14 − 7 = 7

C 18 { 9, 9 — 9 + 9 = 18; 18 − 9 = 9

D 16 { 8, 8 — 8 + 8 = 16; 16 − 8 = 8

7

12	11	14	15	16	13	14	19
− 6	− 10	− 7	− 10	− 8	− 10	− 7	− 10
6	**1**	**7**	**5**	**8**	**3**	**7**	**9**
18	14	18	12	8	8	16	12
− 9	− 10	− 10	− 6	− 4	− 0	− 8	− 10
9	**4**	**8**	**6**	**4**	**8**	**8**	**2**
12	14	16	16	18	6	6	10
− 6	− 7	− 10	− 8	− 9	− 3	− 1	− 1
6	**7**	**6**	**8**	**9**	**3**	**5**	**9**

8

A $2\overset{0}{\cancel{1}}{}^{1}7$ **B** $7\overset{1}{\cancel{2}}{}^{1}04$ **C** $3\overset{0}{\cancel{1}}{}^{1}2$ **D** $\overset{0}{\cancel{1}}{}^{1}730$ **E** $\overset{6}{\cancel{7}}{}^{1}7$ **F** $3\overset{3}{\cancel{4}}{}^{1}81$

9

A	B	C	D	E
348	4325	5347	268	64
− 7	− 10	− 31	− 17	− 3
341	**4315**	**5316**	**251**	**61**

10

A	B	C	D	E
643	685	3426	7356	758
− 510	+ 106	+ 2183	− 1045	− 111
133	**791**	**5609**	**6311**	**647**

Lesson 14

Facts + Problems + Bonus = TOTAL

1

A 508 **B** 3740 **C** 2851 **D** 1418 **E** 231 **F** 3266

2

A 18 { 10, 8 — 18 − 10 = 8; 18 − 8 = 10

B 12 { 10, 2 — 12 − 10 = 2; 12 − 2 = 10

C 19 { 10, 9 — 19 − 10 = 9; 19 − 9 = 10

D 14 { 10, 4 — 14 − 10 = 4; 14 − 4 = 10

3

12	12	12	12	12	12	12	12
− 5	− 3	− 4	− 6	− 3	− 5	− 4	− 5
7	**9**	**8**	**6**	**9**	**7**	**8**	**7**
12	12	12	12	12	12	12	12
− 6	− 4	− 3	− 5	− 6	− 4	− 3	− 5
6	**8**	**9**	**7**	**6**	**8**	**9**	**7**

4

10	7	4	8	6	4	6	3
− 9	− 7	− 0	− 1	− 5	− 4	− 1	− 3
1	**0**	**4**	**7**	**1**	**0**	**5**	**0**
5	8	9	6	4	7	4	5
− 0	− 0	− 8	− 6	− 0	− 1	− 3	− 3
5	**8**	**1**	**0**	**4**	**6**	**1**	**2**

5

16	16	14	14	14	14	18	12
− 10	− 8	− 7	− 10	− 8	− 6	− 9	− 6
6	**8**	**7**	**4**	**6**	**8**	**9**	**6**

Part 5 continues on the next page.

Lesson 14 (continued)

12	16	14	14	12	15	14	10
− 10	− 8	− 6	− 8	− 6	− 10	− 8	− 5
2	**8**	**8**	**6**	**6**	**5**	**6**	**5**
14	14	14	14	8	6	18	16
− 6	− 10	− 7	− 8	− 4	− 3	− 9	− 8
8	**4**	**7**	**6**	**4**	**3**	**9**	**8**

6

A $\overset{6}{\cancel{7}}{}^{1}3$ **B** $9\overset{0}{\cancel{1}}{}^{1}4$ **C** $\overset{2}{\cancel{3}}{}^{1}74$ **D** $\overset{0}{\cancel{1}}{}^{1}380$ **E** $2\overset{0}{\cancel{1}}{}^{1}04$ **F** $5\overset{2}{\cancel{3}}{}^{1}25$

7

A ⧄4□ / − □7□

B ⧄2□ / − □7□

C □□9 / − □□2

D □⧄6□ / − □7□

E □□9 / − □□8

F ⧄3□□ / − □4□□

G □⧄2 / − □3

H □□5□ / − □3□

8

A	B	C	D	E
937	469	6284	4321	5374
− 1	− 8	− 83	− 10	− 62
936	**461**	**6201**	**4311**	**5312**

9

A	B	C	D	E
436	443	734	568	364
+ 216	− 201	− 622	+ 128	− 132
652	**242**	**112**	**696**	**232**

Lesson 15

Test + Facts + Problems + Bonus = TOTAL

1

12	12	12	12	12	12	12	12
− 3	− 6	− 4	− 5	− 6	− 3	− 5	− 4
9	**6**	**8**	**7**	**6**	**9**	**7**	**8**
12	12	12	12	12	12	12	12
− 6	− 3	− 5	− 4	− 3	− 6	− 5	− 4
6	**9**	**7**	**8**	**9**	**6**	**7**	**8**

2

14	14	14	9	7	2	8	6
− 6	− 10	− 8	− 8	− 7	− 0	− 7	− 1
8	**4**	**6**	**1**	**0**	**2**	**1**	**5**
6	9	5	6	14	6	8	14
− 5	− 0	− 4	− 1	− 8	− 6	− 0	− 6
1	**9**	**1**	**5**	**6**	**0**	**8**	**8**
8	5	14	10	4	14	9	7
− 7	− 1	− 8	− 9	− 0	− 6	− 8	− 1
1	**4**	**6**	**1**	**4**	**8**	**1**	**6**

3

A 17 { 10, 7 — 17 − 10 = 7; 17 − 7 = 10

B 19 { 10, 9 — 19 − 10 = 9; 19 − 9 = 10

C 13 { 10, 3 — 13 − 10 = 3; 13 − 3 = 10

D 14 { 10, 4 — 14 − 10 = 4; 14 − 4 = 10

4

A 7214 **B** 9382 **C** 451 **D** 4160 **E** 380 **F** 5246

Lesson 15 (continued)

5

7	7	7	7	7	7	7	7
−1	−3	−2	−0	−3	−2	−1	−3
6	4	5	7	4	5	6	4

6

A. 51~~3~~4 [2 1] B. ~~2~~0 [1 1] C. 6~~3~~0 [2 1] D. ~~7~~82 [6 1] E. 8~~1~~72 [0 1] F. 9~~1~~02 [0 1]

7

A. □☒3□ − □□5□ B. □☒4□ − □□6□ C. □□8 − □□8 D. □☒6□ − □□7□

E. ☒0□□ − □5□□ F. □□3□ − □1□ G. □□2 − □□2 H. □☒7 − □□8

8

12	14	10	12	18	8	6	16
− 6	− 7	− 5	− 6	− 9	−4	−3	− 8
6	7	5	6	9	4	3	8
18	14	14	16	16	12	12	18
−10	−10	− 7	−10	− 8	− 6	−10	− 9
8	4	7	6	8	6	2	9

9

A	B	C	D	E
983	6928	626	2648	46
− 41	− 17	− 13	− 324	− 5
942	6911	613	2324	41

10

A	B	C	D	E
9679	4762	868	4682	4438
−8370	+4610	−467	−2502	+2028
1309	9372	401	2180	6466

Lesson 16

Facts + Problems + Bonus = TOTAL

1

A. 4318 B. 6204 C. 704 D. 1824 E. 314

2

7	7	7	7	7	7	7	7
−3	−1	−2	−0	−2	−3	−1	−3
4	6	5	7	5	4	6	4

3

12	12	12	12	12	12	12	12
− 4	− 5	− 6	− 3	− 5	− 4	− 6	− 3
8	7	6	9	7	8	6	9
12	12	12	12	12	12	12	12
− 5	− 3	− 6	− 4	− 5	− 3	− 6	− 5
7	9	6	8	7	9	6	7

4

16	16	14	14	14	14	12	12
−10	− 8	− 6	−10	− 7	− 8	−10	− 6
6	8	8	4	7	6	2	6
18	14	8	10	10	8	8	8
−10	− 6	−7	− 5	− 9	−4	−1	−8
8	8	1	5	1	4	7	0
9	7	6	7	14	5	8	14
−8	−7	−0	−1	− 8	−0	−7	− 6
1	0	6	6	6	5	1	8

5

A. □☒4 − □□6 B. □□3 − □□0 C. □☒6□ − □□9□ D. □5 − □5

E. ☒6 − □7 F. □□9□ − □□8□ G. ☒0 − □2 H. □□8 − □□7

Lesson 16 (continued)

6

A	B	C	D
[2] ☒[1]4	[7] ☒[1]45	[4] 7☒[1]03	[2] ☒[1]685
− 18	− 371	−6212	−1814
16	474	1291	1871

7

A	B	C	D	E
[4 1] ~~5~~2	[8 1] ~~9~~44	[5 1] 2~~6~~0	[7 1] ~~8~~24	[6 1] 28~~7~~6
−36	−173	−155	−461	−1458
16	771	105	363	1418

8

A	B	C	D	E
387	4258	3686	508	4378
+ 7	− 114	− 513	+ 8	− 68
394	4144	3173	516	4310

F	G	H	I	J
4216	3824	7538	426	653
− 100	+1810	− 430	− 20	+253
4116	5634	7108	406	906

Lesson 17

Facts + Problems + Bonus = TOTAL

1

7	7	7	7	7	7	7	7
−3	−0	−2	−1	−3	−1	−2	−3
4	7	5	6	4	6	5	4

2

A. 3406 B. 182 C. 5414 D. 3811 E. 701

3

A. 12 { 5: 12 − 5 = 7; [7]: 12 − 7 = 5

B. 12 { 4: 12 − 4 = 8; [8]: 12 − 8 = 4

C. 12 { 3: 12 − 3 = 9; [9]: 12 − 9 = 3

D. 12 { 2: 12 − 2 = 10; [10]: 12 − 10 = 2

4

16	16	18	14	18	12	8	6
− 8	−10	− 9	− 7	−10	− 6	−4	−3
8	6	9	7	8	6	4	3
18	8	14	14	16	10	10	16
− 9	−7	− 7	−10	− 8	− 9	− 5	− 8
9	1	7	4	8	1	5	8

5

12	14	12	12	8	12	14	14
− 5	− 6	− 6	− 4	−8	− 3	− 7	− 8
7	8	6	8	0	9	7	6
7	12	14	12	5	9	12	16
−0	− 4	− 6	− 5	−4	−0	− 3	− 8
7	8	8	7	1	9	9	8

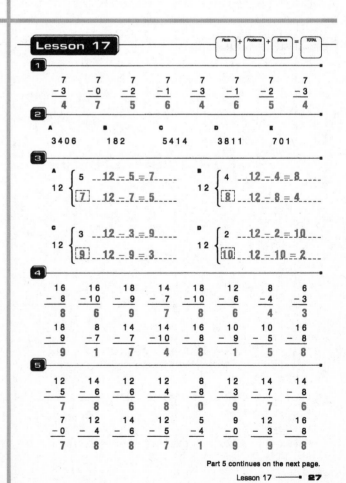

Part 5 continues on the next page.

Lesson 17 (continued)

8	12	18	10	12	18	14	12
−7	−5	−10	−1	−3	−9	−8	−4
1	**7**	**8**	**9**	**9**	**9**	**6**	**8**

6

	A	B	C	D
	□☒5□	□□5□	□8□□	□5
	−□□7□	−□□5□	−□7□□	−□0

	E	F	G	H
	☒0□	□□7	□☒3	□4□□
	−□8□	−□□7	−□□8	−□0□□

7

A	B	C	D
62	6534	4385	3453
−16	−3417	−1195	−1641
46	**3117**	**3190**	**1812**

8

A	B	C	D	E
42	704	6284	3648	2863
−26	−153	−5017	−1807	−1681
16	**551**	**1267**	**1841**	**1182**

9

A	B	C	D	E
749	2371	86	5714	3428
+109	− 70	+26	−1602	− 24
858	**2301**	**112**	**4112**	**3404**

F	G	H	I	J
3864	5624	408	6298	4582
− 732	+ 635	− 4	− 88	+ 182
3132	**6259**	**404**	**6210**	**4764**

Lesson 18

Facts + Problems + Bonus = TOTAL

1

12	7	7	12	12	7	12	12
−4	−2	−3	−3	−5	−1	−3	−4
8	**5**	**4**	**9**	**7**	**6**	**9**	**8**
7	12	7	12	7	12	12	12
−0	−5	−2	−3	−3	−4	−5	−3
7	**7**	**5**	**9**	**4**	**8**	**7**	**9**

2

A. 12 { 4, [8] — **12 − 4 = 8**; **12 − 8 = 4**

B. 12 { 3, [9] — **12 − 3 = 9**; **12 − 9 = 3**

C. 12 { 5, [7] — **12 − 5 = 7**; **12 − 7 = 5**

D. 12 { 2, [10] — **12 − 2 = 10**; **12 − 10 = 2**

3

12	12	12	12	12	12	12	12
−7	−9	−8	−6	−8	−9	−7	−6
5	**3**	**4**	**6**	**4**	**3**	**5**	**6**
12	12	12	12	12	12	12	12
−4	−8	−5	−7	−3	−9	−6	−5
8	**4**	**7**	**5**	**9**	**3**	**6**	**7**

4

A	B	C	D	E
5803	211	6414	814	9909

5

14	14	14	14	12	12	12	12
−6	−10	−8	−7	−10	−5	−4	−6
8	**4**	**6**	**7**	**2**	**7**	**8**	**6**

Part 5 continues on the next page.

Lesson 18 (continued)

18	9	14	16	6	15	16	8
−9	−0	−8	−10	−6	−10	−8	−7
9	**9**	**6**	**6**	**0**	**5**	**8**	**1**
12	10	8	14	3	18	12	8
−10	−1	−4	−6	−0	−9	−6	−8
2	**9**	**4**	**8**	**3**	**9**	**6**	**0**
14	14	5	13	5	16	12	14
−8	−7	−0	−10	−5	−8	−4	−10
6	**7**	**5**	**3**	**0**	**8**	**8**	**4**

6

A	B	C	D	E
396	607	848	942	7875
−248	−497	−670	−180	−2965
148	**110**	**178**	**762**	**4910**

F	G	H	I	J
932	3643	5990	825	3046
−826	−1840	−4189	−160	−1506
106	**1803**	**1801**	**665**	**1540**

7

A	B	C	D	E
8425	4186	386	57	386
+ 35	− 143	+180	+37	− 43
8460	**4043**	**566**	**94**	**343**

F	G	H	I	J
4163	865	391	3286	528
−2033	−455	+ 98	− 43	+ 8
2130	**410**	**489**	**3243**	**536**

Lesson 19

Facts + Problems + Bonus = TOTAL

1

7	7	12	7	12	12	7	12
−3	−0	−3	−1	−5	−4	−2	−3
4	**7**	**9**	**6**	**7**	**8**	**5**	**9**
7	12	7	12	7	12	12	12
−3	−4	−1	−5	−2	−5	−3	−4
4	**8**	**6**	**7**	**5**	**7**	**9**	**8**

2

A. 12 { 4, [8] — **12 − 4 = 8**; **12 − 8 = 4**

B. 12 { 3, [9] — **12 − 3 = 9**; **12 − 9 = 3**

C. 12 { 5, [7] — **12 − 5 = 7**; **12 − 7 = 5**

D. 12 { 2, [10] — **12 − 2 = 10**; **12 − 10 = 2**

3

12	12	12	12	12	12	12	12
−8	−6	−9	−7	−9	−7	−8	−6
4	**6**	**3**	**5**	**3**	**5**	**4**	**6**
12	12	12	12	12	12	12	12
−5	−9	−4	−3	−7	−8	−5	−9
7	**3**	**8**	**9**	**5**	**4**	**7**	**3**

4

14	8	14	9	14	12	18	16
−6	−4	−10	−8	−8	−6	−9	−10
8	**4**	**4**	**1**	**6**	**6**	**9**	**6**
10	14	6	18	14	15	9	17
−5	−6	−0	−9	−8	−10	−9	−10
5	**8**	**6**	**9**	**6**	**5**	**0**	**7**

Part 4 continues on the next page.

14 − 7 = 7	14 − 8 = 6	18 − 9 = 9	6 − 5 = 1	11 − 10 = 1	6 − 3 = 3	14 − 6 = 8	14 − 7 = 7
18 − 10 = 8	18 − 9 = 9	13 − 10 = 3	8 − 4 = 4	15 − 10 = 5	6 − 5 = 1	16 − 8 = 8	13 − 10 = 3

5

A	B	C	D	E
3418	708	9210	5300	121

6

	Borrow marks	Problem	Answer
A	8_1	492 − 286	206
B	8_1	903 − 791	112
C	5_1	681 − 491	190
D	6_1	7043 − 5110	1933
E	6_1	76 − 38	38
F	4_1	3500 − 1350	2150
G	5_1	9601 − 490	9111
H	5_1	6824 − 1904	4920
I	7_1	584 − 466	118
J	6_1	745 − 80	665

7

	Problem	Answer
A	8794 − 1391	7403
B	67 − 36	31
C	3625 + 444	4069
D	646 + 108	754
E	6747 − 4601	2146
F	3898 − 7	3891
G	258 + 8	266
H	9148 − 28	9120
I	3794 + 96	3890
J	5234 − 10	5224

Lesson 20

Facts + Problems + Bonus = TOTAL

1

10 − 5 = 5	10 − 4 = 6	10 − 0 = 10	10 − 1 = 9	10 − 3 = 7	10 − 2 = 8	10 − 4 = 6	10 − 2 = 8
10 − 3 = 7	10 − 4 = 6	10 − 1 = 9	10 − 5 = 5	10 − 2 = 8	10 − 3 = 7	10 − 2 = 8	10 − 4 = 6

2

A. 12 { 4: 12 − 4 = 8; [8]: 12 − 8 = 4 }

B. 12 { 2: 12 − 2 = 10; [10]: 12 − 10 = 2 }

C. 12 { 3: 12 − 3 = 9; [9]: 12 − 9 = 3 }

D. 12 { 5: 12 − 5 = 7; [7]: 12 − 7 = 5 }

3

12 − 5 = 7	12 − 8 = 4	12 − 7 = 5	12 − 9 = 3	12 − 6 = 6	12 − 7 = 5	12 − 8 = 4	12 − 9 = 3
12 − 4 = 8	12 − 9 = 3	12 − 5 = 7	12 − 8 = 4	12 − 3 = 9	12 − 7 = 5	12 − 6 = 6	12 − 10 = 2

4

7 − 3 = 4	7 − 2 = 5	14 − 6 = 8	8 − 0 = 8	14 − 7 = 7	5 − 5 = 0	7 − 3 = 4	14 − 8 = 6
10 − 1 = 9	18 − 9 = 9	6 − 6 = 0	18 − 10 = 8	14 − 8 = 6	12 − 10 = 2	16 − 8 = 8	9 − 8 = 1

Part 4 continues on the next page.

4 − 0 = 4	14 − 6 = 8	8 − 4 = 4	6 − 0 = 6	17 − 10 = 7	18 − 9 = 9	10 − 9 = 1	7 − 3 = 4

5

A	B	C	D	E
3204	814	1941	8111	414

6

	Borrow marks	Problem	Answer
A	2_1 3_1	3248 − 1639	1609
B	4_1 7_1	5082 − 1966	3116
C	8_1 2_1	9436 − 1728	7708
D	2_1 0_1	3410 − 1609	1801

7

	Borrow marks	Problem	Answer
A	7_1	824 − 144	680
B	3_1	7246 − 138	7108
C		384 + 42	426
D	8_1	90 − 49	41
E	5_1	4603 − 50	4553
F		4327 + 1755	6082
G	2_1	830 − 105	725
H	3_1	4681 − 2871	1810
I		3415 + 95	3510
J	1_1	7284 − 92	7192
K	2_1	3248 − 540	2708
L	6_1	5700 − 10	5690
M		480 + 85	565
N	5_1	648 − 180	468
O		286 − 36	250

Lesson 21

Test + Facts + Problems + Bonus = TOTAL

1

12 − 9 = 3	12 − 5 = 7	12 − 6 = 6	12 − 8 = 4	12 − 7 = 5	12 − 4 = 8	12 − 3 = 9	12 − 10 = 2
12 − 8 = 4	12 − 4 = 8	12 − 5 = 7	12 − 7 = 5	12 − 3 = 9	12 − 9 = 3	12 − 7 = 5	12 − 4 = 8

2

A. 7 { 2: 7 − 2 = 5; [5]: 7 − 5 = 2 }

B. 7 { 0: 7 − 0 = 7; [7]: 7 − 7 = 0 }

C. 7 { 3: 7 − 3 = 4; [4]: 7 − 4 = 3 }

D. 7 { 1: 7 − 1 = 6; [6]: 7 − 6 = 1 }

3

14 − 6 = 8	19 − 10 = 9	16 − 8 = 8	16 − 10 = 6	14 − 8 = 6	18 − 9 = 9	12 − 10 = 2	10 − 5 = 5
6 − 6 = 0	18 − 9 = 9	10 − 9 = 1	14 − 7 = 7	16 − 10 = 6	11 − 10 = 1	10 − 1 = 9	8 − 7 = 1

4

7 − 4 = 3	7 − 5 = 2	7 − 6 = 1	7 − 2 = 5	7 − 4 = 3	7 − 3 = 4	7 − 5 = 2	7 − 7 = 0

5

	Problem	Answer
A	435 − 414	21
B	629 − 511	118
C	5734 − 5124	610
D	7765 − 7160	605
E	867 − 450	417

Lesson 21 (continued)

6

A 3400 B 340 C 1824 D 4777 E 477

7

10 − 4 = 6 | 10 − 3 = 7 | 10 − 1 = 9 | 10 − 2 = 8 | 10 − 5 = 5 | 10 − 3 = 7 | 10 − 4 = 6

8

A 50 − 1 = 49 B 30 − 1 = 29 C 80 − 1 = 79
D 90 − 1 = 89 E 40 − 1 = 39 F 20 − 1 = 19
G 60 − 1 = 59 H 70 − 1 = 69

9

A 3420 − 1801 = 1619 B 8486 − 3768 = 4718 C 3092 − 1146 = 1946 D 3240 − 1739 = 1501

10

A 752 − 24 = 728 B 3493 − 610 = 2883 C 3480 − 51 = 3429 D 70 + 38 = 108 E 2400 − 90 = 2310

F 3528 + 528 = 4056 G 738 + 8 = 746 H 9248 − 1068 = 8180 I 7251 − 601 = 6650 J 3846 + 3218 = 7064

K 460 − 355 = 105 L 5284 + 6 = 5290 M 9328 − 16 = 9312 N 4822 − 142 = 4680 O 8605 − 7150 = 1455

Lesson 22

Test + Facts + Problems + Bonus = TOTAL

1

12 − 7 = 5 | 12 − 9 = 3 | 12 − 6 = 6 | 12 − 8 = 4 | 12 − 5 = 7 | 12 − 3 = 9 | 12 − 8 = 4 | 12 − 4 = 8

12 − 7 = 5 | 12 − 5 = 7 | 12 − 4 = 8 | 12 − 8 = 4 | 12 − 6 = 6 | 12 − 9 = 3 | 12 − 3 = 9 | 12 − 10 = 2

2

A 7 { 3: 7 − 3 = 4; [4]: 7 − 4 = 3 }
B 7 { 2: 7 − 2 = 5; [5]: 7 − 5 = 2 }
C 7 { 0: 7 − 0 = 7; [7]: 7 − 7 = 0 }
D 7 { 1: 7 − 1 = 6; [6]: 7 − 6 = 1 }

3

14 − 10 = 4 | 14 − 7 = 7 | 14 − 8 = 6 | 18 − 9 = 9 | 6 − 6 = 0 | 13 − 10 = 3 | 8 − 0 = 8 | 14 − 6 = 8

11 − 10 = 1 | 18 − 10 = 8 | 18 − 9 = 9 | 15 − 10 = 5 | 16 − 10 = 6 | 14 − 6 = 8 | 16 − 8 = 8 | 14 − 8 = 6

4

A 3261 B 5208 C 1418 D 555 E 5555

5

10 − 4 = 6 | 10 − 3 = 7 | 10 − 2 = 8 | 10 − 5 = 5 | 10 − 1 = 9 | 10 − 4 = 6 | 10 − 2 = 8 | 10 − 3 = 7

Lesson 22 (continued)

6

A 438 − 414 = 24 B 329 − 211 = 118 C 4834 − 4124 = 710 D 7968 − 7160 = 808 E 867 − 110 = 757

7

A 60 − 1 = 59 B 70 − 1 = 69 C 20 − 1 = 19
D 40 − 1 = 39 E 30 − 1 = 29 F 50 − 1 = 49
G 90 − 1 = 89 H 80 − 1 = 79

8

A 3062 − 954 = 2108 B 3472 − 1855 = 1617 C 4032 − 1515 = 2517 D 3864 − 956 = 2908

9

A 3908 − 1718 = 2190 B 840 − 280 = 560 C 596 − 490 = 106 D 70 − 35 = 35

E 5802 − 91 = 5711 F 4038 − 510 = 3528 G 3472 − 1455 = 2017 H 785 − 640 = 145

Lesson 23

Facts + Problems + Bonus = TOTAL

1

12 − 9 = 3 | 12 − 4 = 8 | 12 − 7 = 5 | 12 − 8 = 4 | 12 − 6 = 6 | 12 − 3 = 9 | 12 − 5 = 7 | 12 − 4 = 8

12 − 8 = 4 | 12 − 6 = 6 | 12 − 9 = 3 | 12 − 7 = 5 | 12 − 4 = 8 | 12 − 8 = 4 | 12 − 7 = 5 | 12 − 3 = 9

2

A 7 { 3: 7 − 3 = 4; [4]: 7 − 4 = 3 }
B 7 { 1: 7 − 1 = 6; [6]: 7 − 6 = 1 }
C 7 { 0: 7 − 0 = 7; [7]: 7 − 7 = 0 }
D 7 { 2: 7 − 2 = 5; [5]: 7 − 5 = 2 }

3

10 − 4 = 6 | 10 − 3 = 7 | 10 − 2 = 8 | 10 − 5 = 5 | 10 − 2 = 8 | 10 − 4 = 6 | 10 − 1 = 9 | 10 − 5 = 5

10 − 4 = 6 | 10 − 2 = 8 | 10 − 1 = 9 | 10 − 3 = 7 | 10 − 2 = 8 | 10 − 4 = 6 | 10 − 5 = 5 | 10 − 0 = 10

4

14 − 6 = 8 | 12 − 10 = 2 | 18 − 9 = 9 | 18 − 10 = 8 | 14 − 8 = 6 | 15 − 10 = 5 | 16 − 8 = 8 | 10 − 5 = 5

4 − 4 = 0 | 18 − 9 = 9 | 13 − 10 = 3 | 8 − 4 = 4 | 11 − 10 = 1 | 8 − 7 = 1 | 6 − 3 = 3 | 8 − 0 = 8

Lesson 23 (continued)

5

7	7	7	7	7	7	7	7
−5	−4	−6	−7	−4	−7	−6	−5
2	3	1	0	3	0	1	2

6

A 4056 B 5207 C 8024 D 2902 E 4908 F 7067

7

A	B	C	D	E
7986	918	4187	5288	9456
−6970	−914	−3106	−5274	−8413
1016	4	1081	14	1043

8

A 50 − 1 = 49 B 80 − 1 = 79 C 70 − 1 = 69
D 20 − 1 = 19 E 60 − 1 = 59 F 40 − 1 = 39
G 30 − 1 = 29 H 90 − 1 = 89

9

A	B	C	D	E
9440	7030	943	9496	80
−4705	−3509	+971	− 686	−25
4735	3521	1914	8810	55

F	G	H	I	J
3484	3420	4286	90	5290
+1086	− 190	− 196	+25	−3605
4570	3230	4090	115	1685

K	L	M	N	O
375	4034	248	4200	380
−340	−1526	+ 8	− 90	+ 80
35	2508	256	4110	460

Lesson 24

Test + Facts + Problems + Bonus = TOTAL

1

10	10	10	10	10	10	10	10
− 4	− 2	− 3	− 1	− 0	− 2	− 4	− 5
6	8	7	9	10	8	6	5
10	10	10	10	10	10	10	10
− 3	− 1	− 4	− 2	− 5	− 3	− 2	− 4
7	9	6	8	5	7	8	6

2

A 7 { 1: 7 − 1 = 6; 6: 7 − 6 = 1 }
B 7 { 3: 7 − 3 = 4; 4: 7 − 4 = 3 }
C 7 { 0: 7 − 0 = 7; 7: 7 − 7 = 0 }
D 7 { 2: 7 − 2 = 5; 5: 7 − 5 = 2 }

3

14	14	5	5	18	6	14	10
− 6	− 7	−0	−4	−10	−0	−10	− 9
8	7	5	1	8	6	4	1
16	13	7	14	7	8	8	15
− 8	−10	−6	− 8	−6	−7	−4	−10
8	3	1	6	1	1	4	5
10	14	18	8	17	8	14	12
− 1	− 6	− 9	−8	−10	−0	− 8	− 6
9	8	9	0	7	8	6	6

4

A 4006 B 5027 C 7004 D 2902 E 4098 F 7007

Lesson 24 (continued)

5

12	12	12	12	12	12	12	12
− 8	− 7	− 4	− 6	−10	− 5	− 9	− 3
4	5	8	6	2	7	3	9
12	12	12	12	12	12	12	12
− 9	−10	− 8	− 7	− 5	− 3	− 4	− 6
3	2	4	5	7	9	8	6

6

A	B	C	D	E
5864	6993	5485	6274	3521
−4824	−6972	−1431	−6201	−2520
1040	21	4054	73	1001

7

A	B	C	D	E
9958	5260	8356	3860	75
−9139	−3619	+ 156	− 45	+70
819	1641	8512	3815	145

F	G	H	I	J
3040	7410	396	6842	463
− 931	+ 609	− 8	− 71	− 80
2109	8019	388	6771	383

K	L	M	N	O
4588	348	9480	3528	780
+ 28	−140	−1776	− 360	+ 80
4616	208	7704	3168	860

8

A 90 − 1 = 89 B 70 − 1 = 69 C 20 − 1 = 19
D 50 − 1 = 49 E 80 − 1 = 79 F 60 − 1 = 59
G 10 − 1 = 9 H 30 − 1 = 29

Lesson 25

Facts + Problems + Bonus = TOTAL

1

10	10	10	10	10	10	10	10
− 4	− 2	− 3	− 1	− 4	− 2	− 3	− 1
6	8	7	9	6	8	7	9

2

12	12	12	7	12	7	12	7
− 5	− 9	− 8	−3	− 7	−4	− 9	−5
7	3	4	4	5	3	3	2
12	12	7	7	12	12	12	12
− 8	− 7	−4	−5	− 9	− 8	− 7	− 6
4	5	3	2	3	4	5	6

3

12	7	12	14	12	14	8	8
− 3	−5	− 4	− 6	− 5	− 8	−4	−7
9	2	8	8	7	6	4	1
16	14	7	14	7	7	12	18
− 8	− 6	−3	− 7	−4	−2	− 5	− 9
8	8	4	7	3	5	7	9

4

A 5009 B 9035 C 8024 D 7406 E 9004 F 8201

5

16	11	14	12	16	13	17	15
− 9	− 9	− 9	− 9	− 9	− 9	− 9	− 9
7	2	5	3	7	4	8	6

6

A	B	C	D	E
502	704	304	606	402
−395	−186	−118	−408	−123
107	518	186	198	279

Lesson 25 (continued)

7

A	6 { 2, [ ] }	6 − 2 = [ ]
B	[ ] { 5, 3 }	5 + 3 = [ ]
C	7 { 5, [ ] }	7 − 5 = [ ]
D	[ ] { 8, 1 }	8 + 1 = [ ]
E	[ ] { 6, 4 }	6 + 4 = [ ]
F	8 { 1, [ ] }	8 − 1 = [ ]

8

A	B	C	D	E
353	698	5746	3926	7254
− 352	− 594	− 5731	− 3746	− 7048
1	**104**	**15**	**80**	**206**

F	G	H	I	J
6432	6289	408	9068	30
− 1725	− 1378	+ 8	− 8949	+ 75
4707	**4911**	**416**	**119**	**105**

K	L	M	N	O
90	7214	5246	9437	740
− 45	+ 86	− 638	− 7607	+ 259
45	**7300**	**4608**	**1830**	**999**

9

A 70 − 1 = **69** B 30 − 1 = **29** C 80 − 1 = **79**
D 20 − 1 = **19** E 60 − 1 = **59** F 90 − 1 = **89**
G 40 − 1 = **39** H 70 − 1 = **69**

Lesson 26

Facts + Problems + Bonus = TOTAL

1

18	15	11	17	13	12	16	11
− 9	− 9	− 9	− 9	− 9	− 9	− 9	− 9
9	**6**	**2**	**8**	**4**	**3**	**7**	**2**

2

16	10	16	16	16	10	16	16
− 7	− 5	− 8	− 9	− 7	− 3	− 10	− 9
9	**5**	**8**	**7**	**9**	**7**	**6**	**7**
10	16	10	16	16	10	16	16
− 4	− 9	− 2	− 7	− 8	− 3	− 9	− 7
6	**7**	**8**	**9**	**8**	**7**	**7**	**9**

3

7	7	7	7	7	7	7	7
− 5	− 4	− 1	− 2	− 3	− 6	− 4	− 7
2	**3**	**6**	**5**	**4**	**1**	**3**	**0**
7	7	7	7	7	7	7	7
− 6	− 3	− 0	− 5	− 2	− 4	− 3	− 1
1	**4**	**7**	**2**	**5**	**3**	**4**	**6**

4

12	14	18	12	6	8	12	4
− 5	− 6	− 9	− 3	− 1	− 0	− 5	− 4
7	**8**	**9**	**9**	**5**	**8**	**7**	**0**
12	12	14	12	6	14	18	12
− 10	− 4	− 10	− 7	− 5	− 8	− 10	− 8
2	**8**	**4**	**5**	**1**	**6**	**8**	**4**

5

A 4064 B 5002 C 9400 D 8006 E 6010 F 8000

Lesson 26 (continued)

6

A	B	C	D	E
502	806	302	802	406
− 314	− 198	− 189	− 716	− 218
188	**608**	**113**	**86**	**188**

7

A	[7] { 4, 3 }	4 + 3 = [7]
B	7 { 2, [5] }	7 − 2 = [5]
C	9 { 8, [1] }	9 − 8 = [1]
D	[9] { 8, 1 }	8 + 1 = [9]
E	14 { 8, [6] }	14 − 8 = [6]
F	[7] { 5, 2 }	5 + 2 = [7]

8

A	B	C	D	E
9924	7453	2532	934	5824
− 9852	− 6800	+ 2427	− 826	− 4904
72	**653**	**4959**	**108**	**920**

F	G	H	I	J
3492	907	8280	480	6032
− 76	+ 57	− 7219	− 421	+ 540
3416	**964**	**1061**	**59**	**6572**

K	L	M	N	O
4210	3812	3846	564	856
+ 606	− 2910	− 3786	− 510	+ 6
4816	**902**	**60**	**54**	**862**

Lesson 27

Facts + Problems + Bonus = TOTAL

1

17	15	12	17	13	16	14	18
− 9	− 9	− 9	− 9	− 9	− 9	− 9	− 9
8	**6**	**3**	**8**	**4**	**7**	**5**	**9**
16	13	12	11	15	14	17	12
− 9	− 9	− 9	− 9	− 9	− 9	− 9	− 9
7	**4**	**3**	**2**	**6**	**5**	**8**	**3**

2

8	16	16	8	16	16	8	8
− 3	− 7	− 10	− 5	− 9	− 7	− 3	− 5
5	**9**	**6**	**3**	**7**	**9**	**5**	**3**
16	16	8	8	16	8	8	8
− 9	− 7	− 3	− 7	− 7	− 5	− 8	− 3
7	**9**	**5**	**1**	**9**	**3**	**0**	**5**

3

10	7	12	12	10	7	12	10
− 4	− 4	− 8	− 3	− 3	− 5	− 7	− 2
6	**3**	**4**	**9**	**7**	**2**	**5**	**8**
12	10	7	12	10	10	12	7
− 4	− 3	− 6	− 8	− 4	− 9	− 4	− 5
8	**7**	**1**	**4**	**6**	**1**	**8**	**2**
12	7	12	10	7	10	12	12
− 3	− 2	− 5	− 3	− 2	− 2	− 3	− 8
9	**5**	**7**	**7**	**5**	**8**	**9**	**4**
10	12	7	7	12	12	16	7
− 5	− 5	− 3	− 7	− 4	− 10	− 8	− 4
5	**7**	**4**	**0**	**8**	**2**	**8**	**3**

4

- A: 12 { 5, [7] } 12 − 5 = [7]
- B: 7 { 6, [1] } 7 − 6 = [1]
- C: [10] { 9, 1 } 9 + 1 = [10]
- D: [10] { 7, 3 } 7 + 3 = [10]
- E: 7 { 2, [5] } 7 − 2 = [5]
- F: [11] { 10, 1 } 10 + 1 = [11]

5

A 4006 B 508 C 300 D 3000 E 502 F 5020

6

- A: 4046 − 2185 = 1861
- B: 5024 − 1853 = 3171
- C: 3704 − 2198 = 1506
- D: 5029 − 3641 = 1388
- E: 904 − 186 = 718

7

- A: 3460 − 2751 = 709
- B: 6224 − 5160 = 1064
- C: 805 + 95 = 900
- D: 420 − 311 = 109

Part 7 continues on the next page.

- E: 948 + 28 = 976
- F: 5854 − 2917 = 2937
- G: 6953 + 812 = 7765
- H: 992 − 945 = 47
- I: 8470 − 815 = 7655
- J: 3480 + 1581 = 5061
- K: 5480 − 4461 = 1019
- L: 3627 − 61 = 3566
- M: 384 + 26 = 410
- N: 4826 + 26 = 4852
- O: 8352 − 4335 = 4017
- P: 9958 − 9139 = 819
- Q: 5260 − 3619 = 1641
- R: 8356 + 156 = 8512
- S: 3860 − 45 = 3815
- T: 75 + 70 = 145

Lesson 28

Facts + Problems + Bonus = TOTAL

1

15 − 9 = 6	12 − 9 = 3	16 − 9 = 7	14 − 9 = 5	17 − 9 = 8	13 − 9 = 4	18 − 9 = 9	11 − 9 = 2
16 − 9 = 7	14 − 9 = 5	17 − 9 = 8	13 − 9 = 4	18 − 9 = 9	12 − 9 = 3	15 − 9 = 6	11 − 9 = 2

2

16 − 7 = 9	8 − 5 = 3	8 − 7 = 1	8 − 3 = 5	16 − 9 = 7	16 − 7 = 9	8 − 3 = 5	8 − 5 = 3
16 − 8 = 8	16 − 7 = 9	8 − 8 = 0	8 − 5 = 3	16 − 9 = 7	8 − 3 = 5	8 − 5 = 3	16 − 7 = 9

3

12 − 8 = 4	7 − 3 = 4	10 − 4 = 6	12 − 4 = 8	10 − 3 = 7	12 − 5 = 7	7 − 5 = 2	12 − 3 = 9
14 − 8 = 6	10 − 2 = 8	18 − 9 = 9	14 − 6 = 8	12 − 7 = 5	14 − 6 = 8	10 − 4 = 6	16 − 8 = 8
12 − 7 = 5	7 − 3 = 4	12 − 4 = 8	7 − 6 = 1	12 − 3 = 9	7 − 4 = 3	12 − 8 = 4	10 − 2 = 8
7 − 5 = 2	12 − 3 = 9	10 − 2 = 8	14 − 8 = 6	12 − 5 = 7	8 − 8 = 0	7 − 7 = 0	12 − 6 = 6

4

- A: 9 { 8, [1] } 9 − 8 = [1]
- B: [14] { 8, 6 } 8 + 6 = [14]
- C: [12] { 3, 9 } 3 + 9 = [12]
- D: 3 { 1, [2] } 3 − 1 = [2]
- E: [12] { 8, 4 } 8 + 4 = [12]
- F: 7 { 5, [2] } 7 − 5 = [2]

5

A 2045 B 2405 C 8001 D 920 E 4054 F 1099

6

- A: 3045 − 165 = 2880
- B: 4029 − 2850 = 1179
- C: 4206 − 3198 = 1008
- D: 502 − 115 = 387
- E: 3506 − 1208 = 2298

7

- A: 3824 − 2917 = 907
- B: 6520 − 5350 = 1170
- C: 5350 − 4324 = 1026
- D: 1956 − 18 = 1938

Part 7 continues on the next page.

Lesson 28 (continued)

E. 5490 − 4472 = 1018	F. 4615 + 605 = 5220	G. 3265 + 15 = 3280	H. 482 − 474 = 8
I. 9432 − 8726 = 706	J. 394 − 306 = 78	K. 5648 − 839 = 4809	L. 482 − 382 = 100
M. 374 + 186 = 560	N. 94 + 80 = 174	O. 702 + 288 = 990	P. 7410 + 609 = 8019
Q. 396 − 8 = 388	R. 6942 − 71 = 6771	S. 463 − 80 = 383	T. 4588 + 28 = 4616

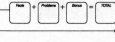

Lesson 29

Facts + Problems + Bonus = TOTAL

1

15 − 9 = 6	13 − 9 = 4	11 − 9 = 2	18 − 9 = 9	16 − 9 = 7	15 − 9 = 6	14 − 9 = 5	12 − 9 = 3
18 − 9 = 9	11 − 9 = 2	14 − 9 = 5	17 − 9 = 8	15 − 9 = 6	12 − 9 = 3	16 − 9 = 7	13 − 9 = 4

2

12 − 7 = 5	10 − 2 = 8	10 − 4 = 6	12 − 8 = 4	7 − 4 = 3	12 − 9 = 3	10 − 3 = 7	7 − 5 = 2
10 − 2 = 8	12 − 3 = 9	10 − 3 = 7	12 − 7 = 5	7 − 4 = 3	12 − 3 = 9	10 − 4 = 6	12 − 6 = 6
12 − 5 = 7	7 − 5 = 2	10 − 4 = 6	14 − 8 = 6	12 − 4 = 8	14 − 6 = 8	12 − 8 = 4	12 − 7 = 5
14 − 6 = 8	12 − 5 = 7	18 − 9 = 9	10 − 3 = 7	12 − 4 = 8	14 − 8 = 6	14 − 7 = 7	12 − 8 = 4

3

16 − 7 = 9	16 − 8 = 8	8 − 3 = 5	16 − 9 = 7	8 − 5 = 3	16 − 7 = 9	8 − 3 = 5	16 − 9 = 7

4

A. 4036 B. 8214 C. 7040 D. 3008 E. 6208

Lesson 29 (continued)

5

A. The big number is 9. A small number is 4. — [9] {[4], [ ]} 9 − 4 = [5]

B. A small number is 10. Another small number is 4. — [ ] {[10], [4]} 10 + 4 = [14]

C. The big number is 5. A small number is 3. — [5] {[3], [ ]} 5 − 3 = [2]

D. A small number is 8. Another small number is 2. — [ ] {[8], [2]} 8 + 2 = [10]

E. A small number is 9. Another small number is 1. — [ ] {[9], [1]} 9 + 1 = [10]

F. The big number is 7. A small number is 3. — [7] {[3], [ ]} 7 − 3 = [4]

6

A. 502 − 185 = 317	B. 308 − 157 = 151	C. 5049 − 768 = 4281	D. 7082 − 3140 = 3942	E. 3068 − 184 = 2884

Lesson 29 (continued)

7

A. 7632 − 808 = 6824	B. 9428 − 50 = 9378	C. 825 + 75 = 900	D. 3420 − 619 = 2801	E. 8425 + 67 = 8492
F. 70 − 45 = 25	G. 850 − 710 = 140	H. 340 + 60 = 400	I. 8250 + 50 = 8300	J. 4215 − 3710 = 505
K. 3040 − 931 = 2109	L. 7410 + 609 = 8019	M. 396 − 8 = 388	N. 6942 − 71 = 6771	O. 463 − 80 = 383
P. 4588 + 28 = 4616	Q. 348 − 140 = 208	R. 9490 − 1776 = 7704	S. 3528 − 360 = 3168	T. 780 + 80 = 860

Lesson 30

Test + Facts + Problems + Bonus = TOTAL

1

12	17	14	16	11	18	15	13
− 9	− 9	− 9	− 9	− 9	− 9	− 9	− 9
3	8	5	7	2	9	6	4
11	17	13	16	12	15	18	14
− 9	− 9	− 9	− 9	− 9	− 9	− 9	− 9
2	8	4	7	3	6	9	5

2

A The big number is 7. A small number is 4.
7 { 4, 3 } 7 − 4 = 3

B A small number is 10. Another small number is 5.
15 { 10, 5 } 10 + 5 = 15

C A small number is 7. Another small number is 3.
10 { 7, 3 } 7 + 3 = 10

D The big number is 10. A small number is 2.
10 { 2, 8 } 10 − 2 = 8

E The big number is 5. A small number is 4.
5 { 4, 1 } 5 − 4 = 1

F A small number is 8. Another small number is 2.
10 { 8, 2 } 8 + 2 = 10

Lesson 30 (continued)

3

12	7	12	12	12	7	10	14
− 8	− 3	− 7	− 4	− 5	− 5	− 3	− 6
4	4	5	8	7	2	7	8
12	12	12	7	12	12	12	7
− 3	− 7	− 5	− 6	− 8	− 4	− 5	− 5
9	5	7	1	4	8	7	2

4

A 7 { 4, 3 } 4 + 3 = 7

B 7 { 2, 5 } 7 − 2 = 5

C 9 { 8, 1 } 9 − 8 = 1

D 9 { 8, 1 } 8 + 1 = 9

E 14 { 8, 6 } 14 − 8 = 6

F 7 { 5, 2 } 5 + 2 = 7

5

16	8	16	8	16	8	16	8
− 7	− 5	− 9	− 3	− 7	− 3	− 7	− 5
9	3	7	5	9	5	9	3

6

A	B	C	D	E
1035	2406	3006	5071	3401

Lesson 30 (continued)

7

A Ann had 9 oranges. She gave 4 oranges to her friends.
9 is a big number.
4 is a small number.

B Jack has 7 books. He buys 2 books.
7 is a small number.
2 is a small number.

C Gloria has 4 pens. She gives away 1 pen.
4 is a big number.
1 is a small number.

D 6 children are in the park. 4 children go home.
6 is a big number.
4 is a small number.

E Roy ate 8 strawberries. Then he ate 5 more.
8 is a small number.
5 is a small number.

F Jane has 5 cats. 3 cats run away.
5 is a big number.
3 is a small number.

8

A	B	C	D	E
802	704	6084	8063	306
− 184	− 553	− 4191	− 6153	− 48
618	151	1893	1910	258

9

A	B	C	D	E
7496	4825	90	70	420
− 1895	+ 1362	+ 45	− 45	− 402
5601	6187	135	25	18

F	G	H	I	J
4826	7428	3245	528	9638
+ 850	− 4357	− 724	+ 418	− 829
5676	3071	2521	946	8809

Lesson 31

Facts + Problems + Bonus = TOTAL

1

A	B	C	D	E
6047	9420	8006	4090	8201

2

14	14	12	12	16	18	18	15
− 9	− 7	− 9	− 10	− 9	− 9	− 10	− 9
5	7	3	2	7	9	8	6
11	13	16	16	15	15	17	11
− 9	− 9	− 9	− 8	− 10	− 9	− 9	− 9
2	4	7	8	5	6	8	2

3

A 10 { 2, 8 } 10 − 2 = 8; 10 − 8 = 2

B 10 { 4, 6 } 10 − 4 = 6; 10 − 6 = 4

C 10 { 3, 7 } 10 − 3 = 7; 10 − 7 = 3

D 10 { 1, 9 } 10 − 1 = 9; 10 − 9 = 1

4

10	10	10	10	10	10	10	10
− 7	− 6	− 9	− 8	− 6	− 9	− 7	− 8
3	4	1	2	4	1	3	2
10	10	10	10	10	10	10	10
− 3	− 7	− 2	− 4	− 8	− 6	− 9	− 1
7	3	8	6	2	4	1	9

Lesson 31 (continued)

5

16	8	12	12	16	7	8	10
− 7	−3	− 7	− 8	− 9	−4	−5	− 2
9	**5**	**5**	**4**	**7**	**3**	**3**	**8**
12	10	12	12	16	12	10	8
− 4	− 2	− 8	− 7	− 7	− 3	− 3	−5
8	**8**	**4**	**5**	**9**	**9**	**7**	**3**
16	10	12	12	14	8	14	12
− 7	− 4	− 9	− 6	− 8	−3	− 6	− 3
9	**6**	**3**	**6**	**6**	**5**	**8**	**9**
12	14	10	14	12	12	12	12
− 7	− 8	− 4	− 6	− 4	− 9	− 5	− 8
5	**6**	**6**	**8**	**8**	**3**	**7**	**4**

6

A
Jill had 7 frogs. She lost 3 frogs.
7 is a **big** number.
3 is a **small** number.

B
Mr. Deloria had 7 books. He bought 3 books.
7 is a **small** number.
3 is a **small** number.

C
Tim had 6 pencils. He found 3 pencils.
6 is a **small** number.
3 is a **small** number.

D
Gino made 5 sandwiches. He gave away 4 sandwiches.
5 is a **big** number.
4 is a **small** number.

7

A
Al found 8 strawberries in the refrigerator. He ate 3 strawberries. How many strawberries are in the refrigerator now?
8 { 3, 5 } **8 − 3 = 5**

Part 7 continues on the next page.

Lesson 31 (continued)

B
Ann has 5 magazines. She buys 4 more magazines. How many magazines does she have now?
9 { 5, 4 } **5 + 4 = 9**

C
Bill had 6 tops. Then he got 2 tops for his birthday. How many tops did Bill have altogether?
8 { 6, 2 } **6 + 2 = 8**

D
Tom had 9 chocolates. He gave 3 chocolates to a friend. How many chocolates does he have left?
9 { 3, 6 } **9 − 3 = 6**

E
Gloria had 7 apples. She ate 3 apples. How many apples does she have left?
7 { 3, 4 } **7 − 3 = 4**

8

- A 504 − 130 = **374** (borrow marks: 4, 1)
- B 702 − 585 = **117** (borrow marks: 6 9, 1)
- C 4064 − 1284 = **2780** (borrow marks: 3 9, 1)
- D 9067 − 7361 = **1706** (borrow marks: 8, 1)
- E 3504 − 328 = **3176** (borrow marks: 4 9, 1)

9

- A 4230 − 3823 = **407** (borrow marks: 3, 1; 2, 1)
- B 4826 + 177 = **5003**
- C 4427 − 624 = **3803** (borrow marks: 3, 1)
- D 50 − 15 = **35** (borrow marks: 4, 1)
- E 785 − 35 = **750**
- F 7265 − 6180 = **1085** (borrow marks: 1, 1)
- G 384 − 376 = **8** (borrow marks: 7, 1)
- H 4280 + 1786 = **6066**
- I 280 + 1280 = **1560**
- J 946 − 870 = **76** (borrow marks: 8, 1)

Lesson 32

Test + Facts + Problems + Bonus = TOTAL

1

5	5	5	5	5	5	5	5
−3	−1	−2	−5	−2	−4	−3	−1
2	**4**	**3**	**0**	**3**	**1**	**2**	**4**

2

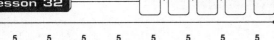

A	5001	4064	2505	340
B	9052	7008	520	3706
C	980	4506	7009	4031
D	6102	870	9003	6041
E	750	8002	5013	4106
F	8309	260	5004	3026

3

14	14	13	11	16	12	12	15
− 9	− 7	− 9	− 9	− 8	− 9	− 5	− 9
5	**7**	**4**	**2**	**8**	**3**	**7**	**6**
18	17	17	10	13	16	12	15
− 9	−10	− 9	− 9	− 9	− 8	− 9	− 9
9	**7**	**8**	**1**	**4**	**8**	**3**	**6**

4

A
10 { 3, 7 } **10 − 3 = 7**; **10 − 7 = 3**

B
10 { 4, 6 } **10 − 4 = 6**; **10 − 6 = 4**

C
10 { 2, 8 } **10 − 2 = 8**; **10 − 8 = 2**

D
10 { 5, 5 } **10 − 5 = 5**; **10 − 5 = 5**

Lesson 32 (continued)

5

10	10	10	10	10	10	10	10
− 8	− 6	− 5	− 7	− 6	− 9	− 8	− 6
2	**4**	**5**	**3**	**4**	**1**	**2**	**4**
10	10	10	10	10	10	10	10
− 2	− 4	− 7	− 3	− 6	− 8	− 1	− 9
8	**6**	**3**	**7**	**4**	**2**	**9**	**1**

6

16	12	10	8	12	12	8	10
− 7	− 8	− 3	−3	− 4	− 7	−5	− 4
9	**4**	**7**	**5**	**8**	**5**	**3**	**6**
7	12	12	16	7	12	8	12
−2	− 6	− 4	− 7	−4	− 8	−5	− 4
5	**6**	**8**	**9**	**3**	**4**	**3**	**8**
14	16	12	8	7	7	12	7
− 6	− 7	− 7	−3	−5	−7	− 5	−3
8	**9**	**5**	**5**	**2**	**0**	**7**	**4**

7

A
Jane had 9 marbles. Her brother gave her 3 more marbles. How many marbles did she have in all?
12 { 9, 3 } **9 + 3 = 12**

B
Sam bought 7 shirts. He lost 3 shirts. How many shirts does he have left?
7 { 3, 4 } **7 − 3 = 4**

C
Joan and Ann baked 4 pies on Friday. On Saturday they baked 5 more pies. How many pies did they bake?
9 { 4, 5 } **4 + 5 = 9**

Part 7 continues on the next page.

Lesson 32 (continued)

D
Bob helped his mother make 7 hamburgers. Then he ate 3 of them. How many hamburgers were left?

[7] {[3], [4]} 7 − 3 = 4

E
You have 8 seashells. A friend gives you 3 more. How many seashells do you have in all?

[11] {[8], [3]} 8 + 3 = 11

8

A	B	C	D	E
532	512	8926	914	5550
− 239	− 18	− 960	− 146	− 5171
293	494	7966	768	379

9

A	B	C	D
4026 − 3156 = 870	3846 + 256 = 4102	8038 − 527 = 7511	406 − 398 = 8

E	F	G	H
705 − 10 = 695	8042 + 160 = 8202	9280 − 8665 = 615	602 − 92 = 510

I	J	K	L
590 − 186 = 404	6804 − 1698 = 5106	40 − 13 = 27	560 − 480 = 80

M	N	O	P
108 + 18 = 126	4026 − 3908 = 118	402 − 6 = 396	9437 − 7607 = 1830

Lesson 33

Test + Facts + Problems + Bonus = TOTAL

1

13	13	13	13	13	13	13	13
− 5	− 10	− 8	− 9	− 5	− 8	− 10	− 5
8	3	5	4	8	5	3	8

2

A 10 {4, [6]} 10 − 4 = 6; 10 − 6 = 4

B 10 {2, [8]} 10 − 2 = 8; 10 − 8 = 2

C 10 {3, [7]} 10 − 3 = 7; 10 − 7 = 3

D 10 {1, [9]} 10 − 1 = 9; 10 − 9 = 1

3

10	10	10	10	10	10	10	10
− 6	− 2	− 4	− 8	− 3	− 7	− 5	− 9
4	8	6	2	7	3	5	1
10	10	10	10	10	10	10	10
− 4	− 2	− 7	− 5	− 8	− 3	− 6	− 4
6	8	3	5	2	7	4	6

4

16	15	8	14	12	16	12	11
− 7	− 9	− 3	− 9	− 7	− 9	− 8	− 9
9	6	5	5	5	7	4	2
12	14	16	12	17	12	8	15
− 7	− 6	− 7	− 3	− 9	− 8	− 3	− 9
5	8	9	9	8	4	5	6

Part 4 continues on the next page.

Lesson 33 (continued)

13	12	8	10	12	14	12	8
− 9	− 6	− 5	− 4	− 4	− 8	− 9	− 3
4	6	3	6	8	6	3	5
15	10	12	16	8	10	14	13
− 9	− 3	− 5	− 7	− 5	− 2	− 7	− 9
6	7	7	9	3	8	7	4

5

5	5	5	5	5	5	5	5
− 2	− 5	− 3	− 0	− 4	− 2	− 1	− 3
3	0	2	5	1	3	4	2

6

A boys, girls, children — children {boys, girls}

B cats, pets, dogs — pets {cats, dogs}

C tools, hammers, saws — tools {hammers, saws}

D stores, houses, buildings — buildings {stores, houses}

E shirts, clothes, dresses — clothes {shirts, dresses}

Lesson 33 (continued)

7

A	B	C	D	E
630	4926	712	6810	474
− 35	− 3749	− 665	− 5119	− 87
595	1177	47	1691	387

8

A
Stefan had 6 card games. He bought 2 more. How many card games did Stefan have in all?

[8] {[6], [2]} 6 + 2 = 8

B
Lisa and a friend are putting up a fence. The fence has 4 gates. Lisa and her friend have finished 3 gates. How many gates do they have left?

[4] {[3], [1]} 4 − 3 = 1

C
Su Lin wrote a story about 12 giraffes. Then Su Lin changed the story. She took out 3 giraffes. How many giraffes did she have left in her story?

[12] {[3], [9]} 12 − 3 = 9

D
Mrs. Tibbs, who plays golf, had 9 golf balls. She got 4 more for her birthday. How many golf balls does she have now?

[13] {[9], [4]} 9 + 4 = 13

E
There are 12 girls playing tennis. 4 go home. How many girls are left?

[12] {[4], [8]} 12 − 4 = 8

F
Our cats had 9 kittens. We gave 2 kittens to a friend. How many kittens did we have left?

[9] {[2], [7]} 9 − 2 = 7

G
Mrs. Rojas wrote 5 children's stories last year. This year she has written 7. How many stories has Mrs. Rojas written in all?

[12] {[5], [7]} 5 + 7 = 12

Lesson 33 (continued)

9

A. 4028 − 3857 = 171	B. 5084 − 2426 = 2658	C. 3460 − 730 = 2730	D. 3820 + 188 = 4008
E. 508 − 27 = 481	F. 3416 + 810 = 4226	G. 5308 − 1248 = 4060	H. 6208 − 5189 = 1019
I. 3040 − 931 = 2109	J. 7410 + 609 = 8019	K. 396 − 8 = 388	L. 6842 − 71 = 6771
M. 4588 + 28 = 4616	N. 348 − 140 = 208	O. 9480 − 1776 = 7704	P. 3528 − 360 = 3168

Lesson 34

Facts + Problems + Bonus = TOTAL

1

9 − 4 = 5	9 − 1 = 8	9 − 3 = 6	9 − 2 = 7	9 − 4 = 5	9 − 3 = 6	9 − 1 = 8	9 − 2 = 7

2

13 − 5 = 8	5 − 2 = 3	13 − 8 = 5	5 − 3 = 2	5 − 2 = 3	13 − 5 = 8	5 − 3 = 2	13 − 8 = 5

3

10 − 7 = 3	10 − 8 = 2	10 − 4 = 6	10 − 2 = 8	10 − 5 = 5	10 − 6 = 4	10 − 3 = 7	10 − 1 = 9
10 − 6 = 4	10 − 9 = 1	10 − 8 = 2	10 − 3 = 7	10 − 4 = 6	10 − 7 = 3	10 − 2 = 8	10 − 3 = 7

4

12 − 9 = 3	16 − 7 = 9	12 − 8 = 4	14 − 9 = 5	12 − 7 = 5	12 − 6 = 6	15 − 9 = 6	11 − 9 = 2
7 − 3 = 4	12 − 8 = 4	16 − 9 = 7	12 − 5 = 7	16 − 7 = 9	12 − 3 = 9	13 − 9 = 4	12 − 4 = 8
8 − 5 = 3	12 − 9 = 3	12 − 4 = 8	16 − 7 = 9	7 − 4 = 3	14 − 6 = 8	18 − 9 = 9	12 − 3 = 9
7 − 5 = 2	12 − 9 = 3	16 − 8 = 8	14 − 8 = 6	11 − 9 = 2	8 − 7 = 1	12 − 4 = 8	8 − 3 = 5

Lesson 34 (continued)

5

A. little boys, big boys, boys — **boys** { little boys / big boys }

B. animals, cows, horses — **animals** { cows / horses }

C. students, boys, girls — **students** { boys / girls }

D. red pens, blue pens, pens — **pens** { red pens / blue pens }

E. toys, balls, blocks — **toys** { balls / blocks }

F. clean cups, cups, dirty cups — **cups** { clean cups / dirty cups }

6

A. Frank wanted to make a big apple pie. He had 6 apples, but they weren't enough. He bought 2 more. How many apples in all did Frank use?
8 { 6 / 2 } 6 + 2 = 8

B. Mrs. Ives is an artist. She made 4 paintings last month. She sold 3 of them. How many paintings did she have left?
4 { 3 / 1 } 4 − 3 = 1

Part 6 continues on the next page.

Lesson 34 (continued)

C. Bob blew up 12 balloons. 3 broke. How many balloons did he have left?
12 { 3 / 9 } 12 − 3 = 9

D. A woman made 9 baskets last week. This week she finished 4 more. How many baskets has she finished so far?
13 { 9 / 4 } 9 + 4 = 13

E. Mr. Yakamura had 12 students in his dance class. 8 students went home. How many students were left?
12 { 8 / 4 } 12 − 8 = 4

F. There were 9 alligators in the river. 2 got out of the river. How many alligators were left in the river?
9 { 2 / 7 } 9 − 2 = 7

G. Marina rode her bike 5 times to a friend's house on Friday. On Saturday she rode 7 times to her friend's house. How many times did Marina ride to her friend's house?
12 { 5 / 7 } 5 + 7 = 12

7

A. 8365 − 1480 = 6885	B. 8080 − 3125 = 4955	C. 4280 + 1786 = 6066	D. 832 + 148 = 980	E. 8140 − 7470 = 670
F. 9038 − 8969 = 69	G. 6024 − 750 = 5274	H. 803 − 40 = 763	I. 7910 − 788 = 7122	J. 8050 − 7868 = 182
K. 702 − 18 = 684	L. 5106 − 1276 = 3830	M. 905 − 125 = 780	N. 380 + 280 = 660	O. 5276 − 4506 = 770

Lesson 35

Facts + Problems + Bonus = TOTAL

1

9	9	9	9	9	9	9	9
− 3	− 1	− 4	− 2	− 0	− 4	− 2	− 3
6	**8**	**5**	**7**	**9**	**5**	**7**	**6**

2

13	16	5	13	5	13	16	13
− 5	− 7	− 3	− 8	− 2	− 5	− 7	− 8
8	**9**	**2**	**5**	**3**	**8**	**9**	**5**

3

A 3 hammers, how many tools, 4 saws

tools [7] { [3] hammers / [4] saws } 4 + 3 = 7

B 3 girls, 5 children, how many boys

children [5] { [3] **girls** / [2] **boys** } **5 − 3 = 2**

C 7 pens, how many red pens, 3 blue pens

pens [7] { [3] **blue pens** / [4] **red pens** } **7 − 3 = 4**

D 5 pets, 4 dogs, how many cats

pets [5] { [4] **dogs** / [1] **cat** } **5 − 4 = 1**

E 5 big boys, how many boys, 3 little boys

boys [8] { [5] **big boys** / [3] **little boys** } **5 + 3 = 8**

Lesson 35 (continued)

4

10	10	10	10	10	10	10	10
− 4	− 9	− 6	− 3	− 5	− 7	− 1	− 8
6	**1**	**4**	**7**	**5**	**3**	**9**	**2**
10	10	10	10	10	10	10	10
− 6	− 2	− 5	− 7	− 9	− 4	− 8	− 1
4	**8**	**5**	**3**	**1**	**6**	**2**	**9**

5

16	14	12	14	12	14	12	8
− 7	− 9	− 5	− 8	− 6	− 6	− 7	− 3
9	**5**	**7**	**6**	**6**	**8**	**5**	**5**
14	7	8	7	13	16	12	14
− 9	− 3	− 5	− 5	− 9	− 7	− 4	− 6
5	**4**	**3**	**2**	**4**	**9**	**8**	**8**
14	16	7	16	11	14	17	8
− 8	− 9	− 4	− 7	− 9	− 7	− 9	− 3
6	**7**	**3**	**9**	**2**	**7**	**8**	**5**
10	8	10	13	16	12	11	15
− 2	− 5	− 4	− 9	− 8	− 9	− 9	− 9
8	**3**	**6**	**4**	**8**	**3**	**2**	**6**

6

A
Dr. Holzburg examined 7 people in their hospital rooms. Then she treated 3 more people in the emergency room. How many people did Dr. Holzburg take care of at the hospital?

[10] { [7] / [3] } **7 + 3 = 10**

B
Pierre and his father built 8 birdhouses. They sold 3 of them. How many birdhouses do they have left?

[8] { [3] / [5] } **8 − 3 = 5**

Part 6 continues on the next page.

Lesson 35 (continued)

C
Mrs. Martin is a bus driver. There are 5 people on her bus. She picks up 5 people. How many people are on the bus now?

[10] { [5] / [5] } **5 + 5 = 10**

D
Audrey took 6 photographs. Then she took 4 more. How many photographs has she taken now?

[10] { [6] / [4] } **6 + 4 = 10**

E
Samantha wrote a play. 7 friends read her play. 5 more friends read her play. How many friends in all read Samantha's play?

[12] { [7] / [5] } **7 + 5 = 12**

F
Barbara's basketball team played 13 games at their school. They played 6 games at other schools. How many games did they play in all?

[19] { [13] / [6] } **13 + 6 = 19**

7

A	B	C	D	E
956	6280	8042	9520	802
− 368	+ 2084	+ 160	− 9480	− 315
588	**8364**	**8202**	**40**	**487**

F	G	H	I	J
5328	3412	6252	6204	904
− 718	− 807	− 6188	+ 1804	− 20
4610	**2605**	**64**	**8008**	**884**

K	L	M	N	O
72	7062	7236	9307	3642
− 14	− 3258	+ 836	− 3692	− 802
58	**3804**	**8072**	**5615**	**2840**

Lesson 36

Facts + Problems + Bonus = TOTAL

1

9	9	9	9	9	9	9	9
− 4	− 2	− 3	− 4	− 1	− 3	− 2	− 2
5	**7**	**6**	**5**	**8**	**6**	**7**	**7**

2

13	13	13	13	13	13	13	13
− 9	− 5	− 7	− 8	− 5	− 7	− 9	− 8
4	**8**	**6**	**5**	**8**	**6**	**4**	**5**

3

A	B	C	D	E
279	146	657	328	864
35	**92**	**8**	**5**	**47**

4

A 8 houses, how many buildings, 2 stores

buildings [10] { [8] **houses** / [2] **stores** } **8 + 2 = 10**

B 5 dogs, 9 pets, how many cats

pets [9] { [5] **dogs** / [4] **cats** } **9 − 5 = 4**

C how many chairs, 2 wooden chairs, 7 plastic chairs

chairs [9] { [2] **wooden chairs** / [7] **plastic chairs** } **2 + 7 = 9**

D 10 tools, 3 saws, how many hammers

tools [10] { [3] **saws** / [7] **hammers** } **10 − 3 = 7**

Part 4 continues on the next page.

E 4 girls, how many boys, 7 children

children [7] { [4] girls / [3] boys — 7 − 4 = 3

F 5 happy children, 3 sad children, how many children

children [8] { [5] happy children / [3] sad children — 5 + 3 = 8

5

10	16	5	10	14	5	10	16
− 7	− 7	− 2	− 6	− 9	− 3	− 8	− 9
3	9	3	4	5	2	2	7
10	11	10	14	10	12	12	12
− 5	− 9	− 4	− 9	− 3	− 8	− 9	− 3
5	2	6	5	7	4	3	9
12	10	18	10	13	12	10	7
− 6	− 2	− 9	− 6	− 9	− 7	− 7	− 3
6	8	9	4	4	5	3	4
12	15	10	12	10	12	17	8
− 4	− 9	− 1	− 3	− 8	− 5	− 9	− 3
8	6	9	9	2	7	8	5

6

A Mrs. Benally bought 8 tickets to visit a museum. She used 2 tickets. How many tickets were left?

[6] tickets — 8 − 2 = 6

B Mr. Harjo had 5 eggs. He used 3 of them to make a cake. How many eggs does Mr. Harjo have now?

[2] eggs — 5 − 3 = 2

Part 6 continues on the next page.

C Rita had 2 gold coins. Then she received 3 more gold coins for her birthday. How many gold coins did Rita end up with?

[5] gold coins — 2 + 3 = 5

D There were 7 pine trees in back of our house. Then we planted 4 more trees in back of the house. How many trees did we have in all?

[11] trees — 7 + 4 = 11

E There are 4 persons in a library. 3 persons go home. How many persons are left in the library?

[1] person — 4 − 3 = 1

7

	A	B	C	D	E
borrow	$6^1 2_1$	$2_1\ 5_1$			$7_1\ 7_1$
	7807	3260	3846	4826	8084
	− 913	− 2752	+ 256	+ 88	− 4557
	6394	508	4102	4914	3527

	F	G	H	I	J
borrow	$6^1 0_1$	$7\,9_1$	$4^1 7_1$	5_1	$6\,9_1$
	7108	804	7582	604	706
	− 1398	− 316	− 7483	− 20	− 589
	5710	488	99	584	117

	K	L	M	N	O
borrow		4_1	$4^1 0_1$	2_1	5_1
	280	50	5162	3046	60
	+ 280	− 23	− 4180	− 2320	− 35
	560	27	982	726	25

Lesson 37

Test + Facts + Problems + Bonus = TOTAL

1

13	13	13	13	13	13	13	13
− 7	− 5	− 8	− 7	− 8	− 5	− 7	− 9
6	8	5	6	5	8	6	4

2

A 9 { 4 / [5] — 9 − 4 = 5; 9 − 5 = 4

B 9 { 2 / [7] — 9 − 2 = 7; 9 − 7 = 2

C 9 { 3 / [6] — 9 − 3 = 6; 9 − 6 = 3

D 9 { 1 / [8] — 9 − 1 = 8; 9 − 8 = 1

3

9	9	9	9	9	9	9	9
− 4	− 2	− 7	− 3	− 6	− 4	− 5	− 3
5	7	2	6	3	5	4	6
9	9	9	9	9	9	9	9
− 6	− 4	− 8	− 5	− 7	− 3	− 4	− 6
3	5	1	4	2	6	5	3

4

5	16	10	16	5	11	10	14
− 2	− 7	− 6	− 9	− 3	− 9	− 7	− 9
3	9	4	7	2	2	3	5
13	10	7	12	8	17	10	7
− 9	− 3	− 2	− 4	− 3	− 9	− 8	− 5
4	7	5	8	5	8	2	2
7	8	12	15	12	7	18	10
− 4	− 5	− 3	− 9	− 5	− 3	− 9	− 6
3	3	9	6	7	4	9	4

Part 4 continues on the next page.

12	10	12	10	12	10	12	12
− 7	− 4	− 6	− 2	− 8	− 7	− 3	− 9
5	6	6	8	4	3	9	3

5

A	B	C	D	E
524	738	916	437	514
74	128	3	536	24

6

A Yasmin and her brother own 3 dogs and 4 cats. They have trained them to do tricks. How many pets have they trained?

pets [7] { [3] dogs / [4] cats — 3 + 4 = 7

B Rick was building a tree house for his brother and sister. Rick has 9 tools. He has 4 hammers. The rest are saws. How many saws does he have?

tools [9] { [4] hammers / [5] saws — 9 − 4 = 5

C Phyllis has 4 white rabbits. She has 3 brown rabbits. How many rabbits does Phyllis have?

rabbits [7] { [4] white rabbits / [3] brown rabbits — 4 + 3 = 7

D There are 7 children in a swimming pool. There are 4 girls in the pool. How many boys are in the pool?

children [7] { [4] girls / [3] boys — 7 − 4 = 3

Part 6 continues on the next page.

E There are 6 houses on Adams Street. There are 4 shops on Adams Street. How many buildings are on Adams Street?

buildings [10] { [6] houses / [4] shops } 6 + 4 = 10

F We put 9 chairs on the stage for a play. There were 4 rocking chairs. The rest were plain wooden chairs. How many plain wooden chairs were there?

chairs [9] { [4] rocking chairs / [5] wooden chairs } 9 − 4 = 5

7

A Eric knew how to play 7 songs on his guitar. This month he learned 2 more songs. How many songs can Eric play now?

[9] songs — 7 + 2 = 9

B Tim's jacket had 8 buttons. Tim lost 3 of the buttons. How many buttons were left on Tim's jacket?

[5] buttons — 8 − 3 = 5

C There were 9 children playing in the park. 4 children went home. How many children were still in the park?

[5] children — 9 − 4 = 5

D Bill has built 8 model planes. If he builds 3 more model planes, how many planes will he have?

[11] planes — 8 + 3 = 11

8

	Carry	Top	Second	Answer
A	4'29,	5303	−1949	3354
B	5'49,	6502	− 786	5716
C	3'69,	4704	− 817	3887
D	4'29,	5306	−4688	618
E		824	+136	960
F	4,	5675	−1832	3843
G	29,	7306	−6218	1088
H	4'0,	5124	−2354	2770
I		1936	+ 236	2172
J	7,	804	−420	384
K	2,	3248	− 701	2547
L		96	−30	66
M		5362	+1682	7044
N	7,	80	−53	27
O	59,	6040	−2180	3860
P	2,	5308	−1248	4060

Lesson 38

Facts + Problems + Bonus = TOTAL

1

9	9	9	9	9	9	9	9
−6	−7	−5	−4	−7	−5	−3	−6
3	2	4	5	2	4	6	3
9	9	9	9	9	9	9	9
−4	−8	−6	−4	−5	−3	−2	−7
5	1	3	5	4	6	7	2

2

A 13 { 7 13 − 7 = 6 / [6] 13 − 6 = 7 }

B 13 { 9 13 − 9 = 4 / [4] 13 − 4 = 9 }

C 13 { 8 13 − 8 = 5 / [5] 13 − 5 = 8 }

D 13 { 10 13 − 10 = 3 / [3] 13 − 3 = 10 }

3

13	13	13	13	13	13	13	13
− 4	− 8	− 9	− 6	− 7	− 9	− 5	− 4
9	5	4	7	6	4	8	9
13	13	13	13	13	13	13	13
− 6	− 9	− 8	− 5	− 4	− 7	− 9	− 4
7	4	5	8	9	6	4	9

4

16	10	7	16	10	11	7	10
− 7	− 6	−2	− 9	− 7	− 9	−5	− 8
9	4	5	7	3	2	2	2
15	10	12	10	7	10	16	10
− 9	− 3	− 9	− 7	−3	− 2	− 9	− 9
6	7	3	3	4	8	7	1

Part 4 continues on the next page.

14	10	7	10	12	18	8	10
− 7	− 8	−4	− 4	− 6	− 9	−4	− 6
7	2	3	6	6	9	4	4
17	10	6	14	10	8	13	12
− 9	− 5	−3	− 9	− 9	−8	− 9	− 6
8	5	3	5	1	0	4	6

5

A 824 − 13 = 811

824
− 13
811

B 627 − 9 = 618

1,
627
− 9
618

C 302 − 37 = 265

29,
302
− 37
265

6

A Miss Soto is a barber. She gave 7 adult haircuts. She also gave 4 children's haircuts. How many haircuts did Miss Soto give?

haircuts [11] { [7] adult haircuts / [4] children's haircuts } 7 + 4 = 11

B In our school there are 10 classrooms. There are 4 large classrooms. How many small classrooms are there?

classrooms [10] { [4] large classrooms / [6] small classrooms } 10 − 4 = 6

C 10 children are playing tennis in the park. There are 7 boys. How many girls are there?

children [10] { [7] boys / [3] girls } 10 − 7 = 3

Part 6 continues on the next page.

D Jaime was giving a party. On Sunday he invited 10 boys. On Monday he invited 6 girls. How many children did Jaime invite to his party?

children [16] { [10] boys; [6] girls } 10 + 6 = 16

E Jack washed and dried 5 new plates. Then he washed and dried 3 old plates. How many plates did Jack wash and dry?

plates [8] { [5] new plates; [3] old plates } 5 + 3 = 8

F There were 9 birds on a park bench. 4 pigeons flew off the bench. Only crows are left. How many crows are left on the bench?

birds [9] { [4] pigeons; [5] crows } 9 − 4 = 5

7

A 4378 − 3520 = 858	**B** 8306 − 3541 = 4765	**C** 8052 − 78 = 7974	**D** 3280 − 180 = 3100	**E** 802 − 98 = 704
F 3502 − 1786 = 1716	**G** 6280 + 1004 = 7284	**H** 938 + 27 = 965	**I** 4125 − 1285 = 2840	**J** 9028 − 8357 = 671
K 7248 − 1761 = 5487	**L** 3215 − 710 = 2505	**M** 5482 − 4859 = 623	**N** 403 − 33 = 370	**O** 8506 − 798 = 7708

8

A Paul made 7 rugs. He sold 4 rugs at an art fair. How many rugs did he have left?
[3] rugs — 7 − 4 = 3

B Mr. Duval bought a parrot that knew 9 words. Mr. Duval taught the parrot 5 words. How many words could the parrot say then?
[14] words — 9 + 5 = 14

C Mr. Wells raises a bridge so tall boats can pass through. He raised the bridge 9 times on Saturday. He also raised the bridge 9 times on Sunday. How many times did Mr. Wells raise the bridge?
[18] times — 9 + 9 = 18

D After our school picnic, Emily and some friends filled 9 bags with litter from the play area. Then they filled 2 bags with litter from tables. How many bags of litter did Emily and her friends collect?
[11] bags — 9 + 2 = 11

Lesson 39

Facts + Problems + Bonus = TOTAL

1

9 − 6 = 3	9 − 7 = 2	9 − 4 = 5	9 − 5 = 4	9 − 8 = 1	9 − 3 = 6	9 − 6 = 3	9 − 7 = 2
9 − 2 = 7	9 − 0 = 9	9 − 5 = 4	9 − 1 = 8	9 − 6 = 3	9 − 2 = 7	9 − 7 = 2	9 − 4 = 5

2

A 13 { 7; [6] } 13 − 7 = 6; 13 − 6 = 7

B 13 { 9; [4] } 13 − 9 = 4; 13 − 4 = 9

C 13 { 8; [5] } 13 − 8 = 5; 13 − 5 = 8

D 13 { 10; [3] } 13 − 10 = 3; 13 − 3 = 10

3

13 − 4 = 9	13 − 9 = 4	13 − 5 = 8	13 − 8 = 5	13 − 4 = 9	13 − 6 = 7	13 − 7 = 6	13 − 9 = 4
13 − 7 = 6	13 − 4 = 9	13 − 8 = 5	13 − 4 = 9	13 − 5 = 8	13 − 6 = 7	13 − 7 = 6	13 − 8 = 5

4

16 − 7 = 9	16 − 9 = 7	13 − 9 = 4	5 − 2 = 3	11 − 9 = 2	14 − 6 = 8	8 − 3 = 5	14 − 9 = 5
10 − 6 = 4	15 − 9 = 6	10 − 8 = 2	12 − 3 = 9	10 − 6 = 4	17 − 9 = 8	10 − 7 = 3	13 − 9 = 4

Part 4 continues on the next page.

12 − 9 = 3	12 − 5 = 7	10 − 8 = 2	18 − 9 = 9	10 − 4 = 6	10 − 8 = 2	12 − 3 = 9	12 − 8 = 4

5

A 4208 − 60 = 4148 (4208 − 60 = 4148)

B 904 − 7 = 897 (904 − 7 = 897)

C 5154 − 148 = 5006 (5154 − 148 = 5006)

6

A Ivan went to the market and bought 6 carrots. He also bought 4 onions. How many vegetables did Ivan buy?

vegetables [10] { [6] carrots; [4] onions } 6 + 4 = 10

B A science class was studying birds. On a trip to the mountains the students saw 8 birds. 7 were hawks. The rest were eagles. How many eagles did they see?

birds [8] { [7] hawks; [1] eagle } 8 − 7 = 1

C Our team had 7 blue uniforms. The rest of the uniforms were red. There were 9 uniforms in all. How many red uniforms were there?

uniforms [9] { [7] blue uniforms; [2] red uniforms } 9 − 7 = 2

Part 6 continues on the next page.

D Kitty and her friends wash cars on weekends. One Saturday Kitty washed 5 cars. 3 of them were new. How many old cars did she wash?

cars — 5 { 3 new cars; 2 old cars } — 5 − 3 = 2

E Every Sunday 6 people come to the tennis courts. 4 women play tennis on one court. Men play on the other court. How many men play on the other court?

people — 6 { 4 women; 2 men } — 6 − 4 = 2

F Fumiko put out birdseed. He watched as 9 blackbirds flew down to eat. Next 9 robins arrived. How many birds ate seed?

birds — 18 { 9 blackbirds; 9 robins } — 9 + 9 = 18

7

A 828 + 80 = 908	B 6308 − 3889 = 2419	C 9246 − 86 = 9160	D 406 − 355 = 51	E 802 − 376 = 426
F 4082 − 3092 = 990	G 4268 − 608 = 3660	H 3642 + 1082 = 4724	I 3304 − 617 = 2687	J 6047 − 3442 = 2605
K 3826 + 800 = 4626	L 3045 + 2955 = 6000	M 6468 − 2580 = 3888	N 7045 − 6182 = 863	O 4304 − 687 = 3617

8

A There were 10 children at a party. 6 of the children went home early. How many children were still at the party?
4 children — 10 − 6 = 4

B Sophia had 10 model cars. She built 6 more model cars. How many cars does she have now?
16 cars — 10 + 6 = 16

C If you have 7 books and your friend gives you 4 books, how many books will you have?
11 books — 7 + 4 = 11

D Stella had 9 apples. She gave 3 apples to her horse. How many apples does Stella have left?
6 apples — 9 − 3 = 6

Lesson 40

Facts + Problems + Bonus = TOTAL

1

13 − 4 = 9, 13 − 7 = 6, 13 − 6 = 7, 13 − 5 = 8, 13 − 9 = 4, 13 − 7 = 6, 13 − 4 = 9, 13 − 6 = 7

13 − 5 = 8, 13 − 8 = 5, 13 − 4 = 9, 13 − 8 = 5, 13 − 6 = 7, 13 − 9 = 4, 13 − 7 = 6, 13 − 4 = 9

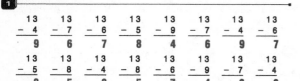

2

A 9 { 3: 9 − 3 = 6; 6: 9 − 6 = 3 }

B 9 { 2: 9 − 2 = 7; 7: 9 − 7 = 2 }

C 9 { 4: 9 − 4 = 5; 5: 9 − 5 = 4 }

D 9 { 1: 9 − 1 = 8; 8: 9 − 8 = 1 }

3

9 − 5 = 4, 9 − 7 = 2, 9 − 6 = 3, 9 − 3 = 6, 9 − 5 = 4, 9 − 4 = 5, 9 − 7 = 2, 9 − 2 = 7

9 − 4 = 5, 9 − 6 = 3, 9 − 8 = 1, 9 − 2 = 7, 9 − 7 = 2, 9 − 1 = 8, 9 − 5 = 4, 9 − 3 = 6

4

10 − 5 = 5, 16 − 7 = 9, 12 − 7 = 5, 10 − 4 = 6, 16 − 9 = 7, 11 − 9 = 2, 10 − 7 = 3, 15 − 9 = 6

12 − 4 = 8, 13 − 9 = 4, 14 − 7 = 7, 12 − 3 = 9, 10 − 6 = 4, 17 − 9 = 8, 8 − 3 = 5, 10 − 3 = 7

Part 4 continues on the next page.

10 − 2 = 8, 10 − 5 = 5, 18 − 9 = 9, 10 − 2 = 8, 16 − 7 = 9, 14 − 9 = 5, 10 − 8 = 2, 10 − 4 = 6

7 − 5 = 2, 10 − 3 = 7, 12 − 3 = 9, 12 − 5 = 7, 10 − 9 = 1, 8 − 5 = 3, 12 − 8 = 4, 7 − 3 = 4

5

A The volleyball team had 7 children. 4 were girls. How many were boys?
3 boys — 7 − 4 = 3

B Jorge wrote 8 songs in all last winter. He wrote 7 sad songs. The rest were funny songs. How many funny songs did he write?
1 funny song — 8 − 7 = 1

C At a country fair, Joan sold 10 peaches. Then she sold 5 pears. How many pieces of fruit did Joan sell?
15 pieces of fruit — 10 + 5 = 15

D Mrs. Stein owns a yarn shop. Last winter she bought 7 big boxes of yarn. She also bought 4 small boxes of yarn. How many boxes of yarn did she buy?
11 boxes — 7 + 4 = 11

E In the park, 4 men were jogging. There were 9 people jogging in the park. How many women were jogging?
5 women — 9 − 4 = 5

F Pam collected shells at the beach. She found 4 gray shells on Monday. On Tuesday she found 10 white shells. How many shells in all did Pam find?
14 shells — 4 + 10 = 14

Lesson 40 (continued)

6

A There were 532 children in our school. 18 children moved to another school. How many children are in our school now?

532 − 18 = 514

514 children

B Mr. Ramirez is a plant scientist. Last year he found 2431 plants. This year he found 92 more plants. How many plants does Mr. Ramirez have now?

2431 + 92 = 2523

2523 plants

C Miss Tanaka owns a gas station. She had 8240 liters of gas. She sold 6105 liters of gas. How many liters does she have now?

8240 − 6105 = 2135

2135 liters

7

A 802 − 796 = **6**	B 69 + 60 = **129**	C 6066 − 3919 = **2147**	D 3502 − 786 = **2716**	E 9654 + 423 = **10,077**
F 430 − 284 = **146**	G 8345 + 1245 = **9590**	H 7042 − 3638 = **3404**	I 4036 − 81 = **3955**	J 4130 − 9 = **4121**
K 5103 − 4941 = **162**	L 4506 − 798 = **3708**	M 4310 − 900 = **3410**	N 3042 − 1998 = **1044**	O 486 + 16 = **502**

Lesson 41

Facts + Problems + Bonus = TOTAL

1

9 − 7 = **2**	9 − 4 = **5**	9 − 3 = **6**	9 − 6 = **3**	9 − 2 = **7**	9 − 8 = **1**	9 − 7 = **2**	9 − 1 = **8**
9 − 5 = **4**	9 − 6 = **3**	9 − 4 = **5**	9 − 7 = **2**	9 − 3 = **6**	9 − 6 = **3**	9 − 2 = **7**	9 − 7 = **2**

2

A 13 { 5: 13 − 5 = 8; [8]: 13 − 8 = 5

B 13 { 6: 13 − 6 = 7; [7]: 13 − 7 = 6

C 13 { 9: 13 − 9 = 4; [4]: 13 − 4 = 9

D 13 { 10: 13 − 10 = 3; [3]: 13 − 3 = 10

3

13 − 7 = **6**	13 − 9 = **4**	13 − 4 = **9**	13 − 8 = **5**	13 − 5 = **8**	13 − 6 = **7**	13 − 4 = **9**	13 − 7 = **6**
13 − 6 = **7**	13 − 8 = **5**	13 − 4 = **9**	13 − 5 = **8**	13 − 9 = **4**	13 − 7 = **6**	13 − 6 = **7**	13 − 8 = **5**

4

10 − 6 = **4**	16 − 9 = **7**	10 − 7 = **3**	14 − 9 = **5**	10 − 4 = **6**	16 − 7 = **9**	10 − 3 = **7**	11 − 9 = **2**
8 − 3 = **5**	10 − 6 = **4**	12 − 3 = **9**	7 − 3 = **4**	12 − 4 = **8**	16 − 9 = **7**	10 − 2 = **8**	17 − 9 = **8**

Part 4 continues on the next page.

Lesson 41 (continued)

10 − 8 = **2**	14 − 9 = **5**	10 − 3 = **7**	8 − 5 = **3**	15 − 9 = **6**	7 − 5 = **2**	12 − 3 = **9**	12 − 7 = **5**
12 − 3 = **9**	14 − 6 = **8**	16 − 8 = **8**	10 − 3 = **7**	7 − 2 = **5**	10 − 7 = **3**	14 − 8 = **6**	7 − 5 = **2**

5

A 5102 − 87 = **5015**	B 6104 − 268 = **5836**	C 3106 − 418 = **2688**	D 7104 − 237 = **6867**

6

A Our school had a dog show for German shepherds and hunting dogs. There were 15 German shepherds in the show. 19 dogs had been brought to the show. How many hunting dogs were there?

19 − 15 = 4

4 hunting dogs

B In the morning 16 jet planes left New York. In the afternoon 74 small planes left New York. How many planes in all left New York?

16 + 74 = 90

90 planes

C There are 7 rosebushes in our garden. We have 4 red rosebushes. The rest are yellow. How many yellow rosebushes do we have?

7 − 4 = 3

3 yellow rosebushes

D There are 14 clean cups on the shelf. There are 19 dirty cups in the sink. How many cups are there in all?

14 + 19 = 33

33 cups

Part 6 continues on the next page.

Lesson 41 (continued)

E 14 women were in the play. 29 people were in the play. How many men were in the play?

29 − 14 = 15

15 men

F My aunt has 28 white flowers. She has 39 flowers in all. The rest of the flowers are pink. How many pink flowers does my aunt have?

39 − 28 = 11

11 pink flowers

7

A A pet shop had 186 birds. The shop sold 48 birds. How many birds does the shop have now?

186 − 48 = 138

138 birds

B Last month Mr. Vanderpool sold 3184 loaves of bread at his bakery. This month Mr. Vanderpool sold 42 more loaves of bread. How many loaves has he sold?

3184 + 42 = 3226

3226 loaves

8

A 920 − 397 = **523**	B 8506 − 6788 = **1718**	C 3514 − 3439 = **75**	D 8426 + 16 = **8442**	E 608 − 29 = **579**
F 7070 − 4028 = **3042**	G 6059 − 1532 = **4527**	H 4109 − 29 = **4080**	I 3704 − 886 = **2818**	J 3148 + 120 = **3268**

Lesson 42

Facts + Problems + Bonus = TOTAL

1

10	7	11	8	9	7	10	9
− 2	− 2	− 2	− 2	− 2	− 2	− 2	− 2
8	5	9	6	7	5	8	7
8	9	11	10	8	7	11	10
− 2	− 2	− 2	− 2	− 2	− 2	− 2	− 2
6	7	9	8	6	5	9	8

2

13	13	13	13	13	13	13	13
− 4	− 9	− 7	− 4	− 5	− 8	− 6	− 10
9	4	6	9	8	5	7	3
13	13	13	13	13	13	13	13
− 8	− 4	− 9	− 6	− 5	− 7	− 4	− 6
5	9	4	7	8	6	9	7

3

9	9	14	9	16	10	9	9
− 4	− 7	− 6	− 6	− 7	− 8	− 3	− 5
5	2	8	3	9	2	6	4
10	13	9	14	14	12	9	17
− 4	− 9	− 1	− 8	− 9	− 3	− 4	− 9
6	4	8	6	5	9	5	8
12	9	15	16	11	13	14	9
− 3	− 7	− 9	− 7	− 9	− 9	− 8	− 7
9	2	6	9	2	4	6	2
12	16	12	9	12	14	12	12
− 5	− 7	− 8	− 8	− 4	− 6	− 3	− 9
7	9	4	1	8	8	9	3

4

A	B	C	D
3104	8102	4102	4102
− 68	− 46	− 538	− 586
3036	8056	3564	3516

5

A Bill painted 43 houses. He painted 52 stores. How many buildings did Bill paint?

43 + 52 = 95

95 buildings

B In the waiting room of a doctor's office there is a huge fish tank. There are 14 goldfish in the tank and the rest are sunfish. There are 30 fish in all in the tank. How many sunfish are in the tank?

30 − 14 = 16

16 sunfish

C There are 40 girls in the kindergarten of school. There are 75 children in kindergarten. How many are boys?

75 − 40 = 35

35 boys

D Miss Manos is selling tickets. She sold 31 circus tickets this morning. This afternoon she sold 50 movie tickets. How many tickets in all did she sell?

31 + 50 = 81

81 tickets

E Marcia Kabatie has entered many skating contests. 43 were ice-skating contests and the rest were roller-skating ones. She has taken part in 60 contests in all. How many roller-skating contests has she entered?

60 − 43 = 17

17 roller-skating contests

6

A	B	C	D	E
863	8504	4280	534	9306
+ 64	− 768	− 859	− 237	− 959
927	7736	3421	297	8347

F	G	H	I	J
9076	4280	3408	8241	3504
− 3179	− 365	− 2614	− 848	− 2614
5897	3915	794	7393	890

7

A Tom feeds 185 chickens on his aunt's farm. His aunt got 15 more chickens. How many chickens will Tom have to feed now?

185 + 15 = 200

200 chickens

B 5246 attended a basketball game. 3138 left before the game was over. How many people stayed for the entire game?

5246 − 3138 = 2108

2108 people

Lesson 43

Test + Facts + Problems + Bonus = TOTAL

1

7	8	6	10	8	5	11	9
− 2	− 2	− 2	− 2	− 2	− 2	− 2	− 2
5	6	4	8	6	3	9	7
5	10	6	11	7	8	4	9
− 2	− 2	− 2	− 2	− 2	− 2	− 2	− 2
3	8	4	9	5	6	2	7

2

13	14	9	13	13	13	9	13
− 5	− 9	− 3	− 6	− 9	− 8	− 7	− 4
8	5	6	7	4	5	2	9
16	13	10	9	13	12	14	9
− 7	− 7	− 3	− 5	− 5	− 3	− 6	− 5
9	6	7	4	8	9	8	4
12	13	13	12	16	9	13	10
− 8	− 6	− 8	− 3	− 7	− 4	− 4	− 5
4	7	5	9	9	5	9	5
14	9	8	13	13	8	10	7
− 6	− 8	− 3	− 4	− 7	− 5	− 4	− 3
8	1	5	9	6	3	6	4

3

15	15	15	15	15	15	15	15
− 7	− 8	− 10	− 8	− 9	− 7	− 10	− 8
8	7	5	7	6	8	5	7

4

A Felipe is a bird-watcher for the junior science club. Last fall Felipe saw 15 woodpeckers. He also saw 64 robins. How many birds did Felipe see?

15 + 64 = 79

79 birds

Part 4 continues on the next page.

Lesson 43 (continued)

B 114 boys were in a children's hospital. There were 218 children in the hospital. How many girls were in the hospital?

218 − 114 = 104

104 girls

C A jewelry shop sold 14 silver watches in one month. The rest of the watches that were sold were gold. Customers bought 25 watches that month. How many gold watches did the shop sell?

25 − 14 = 11

11 gold watches

D Mel offered to wash the big windows in his house. His house has 15 small windows. There are 26 windows altogether. How many big windows are there?

26 − 15 = 11

11 big windows

E Tamara finished painting a large building. She used 14 yellow cans of paint. She used 20 blue cans of paint. How many cans of paint did Tamara use in all?

14 + 20 = 34

34 cans of paint

5

A	B	C	D	E
800	900	8500	9700	800
− 154	− 180	− 7145	− 7360	− 142
646	720	1355	2340	658

6

A	B	C	D	E
5106	970	864	4104	4960
− 468	− 734	+ 132	− 27	− 3899
4638	236	996	4077	1061

Part 6 continues on the next page.

Lesson 43 (continued)

F	G	H	I	J
5140	6028	8109	9304	8605
− 98	+ 978	− 239	− 8949	− 554
5042	7006	7870	355	8051

7

A In a wildlife park, rangers put tags on the legs of 4380 birds. Later, the rangers found that 90 of the birds had flown away. How many were left?

4380 − 90 = 4290

4290 birds

B There were 135 children in the swimming pool. 15 children jumped into the pool. How many children were in the pool then?

135 + 15 = 150

150 children

C Sally had 34 school friends. Over the years 19 of her friends have moved away. How many friends are left?

34 − 19 = 15

15 friends

D A store had 1485 coats. It sold 665 coats. How many are left?

1485 − 665 = 820

820 coats

Lesson 44

Test + Facts + Problems + Bonus = TOTAL

1

9 − 2 = 7	6 − 2 = 4	8 − 2 = 6	5 − 2 = 3	10 − 2 = 8	3 − 2 = 1	7 − 2 = 5	11 − 2 = 9
8 − 2 = 6	4 − 2 = 2	9 − 2 = 7	7 − 2 = 5	10 − 2 = 8	5 − 2 = 3	11 − 2 = 9	6 − 2 = 4

2

A 15 { 9, 6 }: 15 − 9 = 6; 15 − 6 = 9

B 15 { 8, 7 }: 15 − 8 = 7; 15 − 7 = 8

3

15 − 6 = 9	15 − 8 = 7	15 − 10 = 5	15 − 9 = 6	15 − 6 = 9	15 − 8 = 7	15 − 7 = 8	15 − 6 = 9
15 − 8 = 7	15 − 7 = 8	15 − 10 = 5	15 − 6 = 9	15 − 8 = 7	15 − 7 = 8	15 − 6 = 9	15 − 8 = 7

4

13 − 5 = 8	16 − 9 = 7	13 − 9 = 4	13 − 6 = 7	9 − 3 = 6	13 − 7 = 6	9 − 5 = 4	13 − 4 = 9
13 − 9 = 4	13 − 8 = 5	15 − 9 = 6	9 − 6 = 3	13 − 4 = 9	9 − 4 = 5	11 − 9 = 2	14 − 6 = 8
14 − 9 = 5	13 − 5 = 8	10 − 3 = 7	17 − 9 = 8	13 − 8 = 5	12 − 9 = 3	9 − 7 = 2	13 − 5 = 8

Part 4 continues on the next page.

Lesson 44 (continued)

9 − 4 = 5	18 − 9 = 9	13 − 4 = 9	16 − 9 = 7	9 − 3 = 6	13 − 8 = 5	12 − 5 = 7	10 − 6 = 4

5

A There are 143 cabins up at the lake. This year more cabins were built. Now there are 160. How many more cabins were built at the lake?

160 − 143 = 17

17 cabins

B My dad's store was having a sale. There were 114 people in the store. Some more people came in. Then there were 142 people in the store. How many more people came into the store?

142 − 114 = 28

28 people

C Benji had a stamp collection. When the year began, he had 1436 stamps. During the year he got 148 stamps. How many stamps did Benji have at the end of the year?

1436 + 148 = 1584

1584 stamps

D A meal was served at a banquet. There were 143 people sitting at the tables. Some more people arrived. Now there are 160 people at the tables. How many more people arrived?

160 − 143 = 17

17 people

E When the game started, there were 842 people watching it. During the game another 58 people began to watch it. How many people were watching the game altogether?

842 + 58 = 900

900 people

Subtraction Answer Key **27**

6

A 800 − 1 = 799 B 600 − 1 = 599 C 400 − 1 = 399

D 100 − 1 = 99 E 700 − 1 = 699 F 200 − 1 = 199

7

A Mrs. Kirk owns a car rental business. She rented 46 sports cars. She rented 64 station wagons. How many cars in all did Mrs. Kirk rent?

46 + 64 = 110

110 cars

B 90 children were taking piano lessons. 27 more children started piano lessons. How many children were taking piano lessons then?

90 + 27 = 117

117 children

C At camp the children hung their bathing suits on clotheslines to dry. This morning there were 32 wet bathing suits on the clothesline. The rest were dry. There were 70 bathing suits in all. How many dry bathing suits were there?

6,70 − 32 = 38

38 dry bathing suits

D A farmer delivered 346 boxes of strawberries to a market. After the market opened, 99 boxes of strawberries were sold in one hour. How many boxes of strawberries does the market still have to sell?

2'3,346 − 99 = 247

247 boxes

Part 7 continues on the next page.

E A flower shop sold 27 bunches of daisies in the morning. The shop sold 30 bunches of roses in the afternoon. How many bunches of flowers in all were sold?

27 + 30 = 57

57 bunches

F Mr. Elkins drives a school bus. Today 27 of his passengers were girls. There were 40 passengers in all. How many boys were there?

3,40 − 27 = 13

13 boys

G There were 146 cars in a parking lot. There were 94 clean cars. How many dirty cars were in the parking lot?

146 − 94 = 52

52 dirty cars

8

A	B	C	D	E
6'39,	3'09,	4'0,		3'09,
7400	4100	5134	705	4106
−2739	− 328	− 994	+288	− 258
4661	3772	4140	993	3848

F	G	H	I	J
3'29,	49'4,		29'5,	7,
4300	5052	3842	3064	800
−3662	−4678	+1586	−1998	−530
638	374	5428	1066	270

Lesson 45

Facts + Problems + Bonus = TOTAL

1

7	11	5	8	9	6	10	7
−2	−2	−2	−2	−2	−2	−2	−2
5	9	3	6	7	4	8	5
9	4	5	10	3	6	11	8
−2	−2	−2	−2	−2	−2	−2	−2
7	2	3	8	1	4	9	6

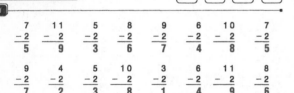

2

A 15 { 9, 6 }: 15 − 9 = 6; 15 − 6 = 9

B 15 { 8, 7 }: 15 − 8 = 7; 15 − 7 = 8

3

15	15	15	15	15	15	15	15
−6	−8	−10	−7	−9	−6	−8	−7
9	7	5	8	6	9	7	8
15	15	15	15	15	15	15	15
−9	−6	−7	−8	−6	−7	−9	−8
6	9	8	7	9	8	6	7

4

13	9	13	13	16	14	13	9
−4	−4	−9	−6	−7	−6	−5	−3
9	5	4	7	9	8	8	6
9	13	11	14	13	9	13	12
−6	−6	−9	−8	−4	−5	−8	−9
3	7	2	6	9	4	5	3
13	13	9	16	13	9	13	15
−8	−4	−6	−7	−7	−3	−5	−9
5	9	3	9	6	6	8	6

Part 4 continues on the next page.

14	9	16	13	10	9	13	10
−9	−4	−9	−7	−3	−5	−9	−4
5	5	7	6	7	4	4	6

5

A When the movie began, there were 146 people in the theater. By the end of the movie there were 206 people in the theater. How many people came into the theater after the movie started?

1,206 − 146 = 60

60 people

B Last summer we took photographs of a building that was being built. At that time the building had 58 floors. In winter the building was finished. It had 70 floors. How many floors were built after we took photographs?

6,70 − 58 = 12

12 floors

C 315 children took the bus to summer camp. Another 46 children took the train to camp. How many children went to summer camp in all?

315 + 46 = 361

361 children

D At a beach there were 54 swimmers. 40 more swimmers came. How many swimmers were at the beach then?

54 + 40 = 94

94 swimmers

E A dress shop had 23 dresses. Some more dresses came in a truck. Now there are 70 dresses in the shop. How many dresses came in the truck?

6,70 − 23 = 47

47 dresses

Lesson 45 (continued)

6

A 800 − 1 = 799 B 400 − 1 = 399 C 100 − 1 = 99

D 600 − 1 = 599 E 900 − 1 = 899 F 200 − 1 = 199

7

A We had a picnic for some friends. 37 children and 40 adults came. How many people came to the picnic?

37 + 40 = 77

77 people

B The big dinner party at Robert's house is over. He has 42 dirty dishes. There are 60 dishes in all. How many clean dishes does Robert have?

60 (borrow: 5, 1) − 42 = 18

18 clean dishes

C Our family had 230 puzzles. We lost 8 puzzles when we moved. Now how many puzzles do we have left?

230 (borrow: 2, 1) − 8 = 222

222 puzzles

D Some lumbermen cut down 1456 tall trees in a forest. They cut down 138 short trees. How many trees in all did the lumbermen cut down?

1456 + 138 = 1594

1594 trees

E A fruit grower delivered fruit to a hotel kitchen. There were 50 boxes of pears and 24 boxes of plums. How many boxes of fruit did the grower deliver to the hotel kitchen?

50 + 24 = 74

74 boxes

Part 7 continues on the next page.

Lesson 45 (continued)

F There were 240 passengers on an airplane. There were 70 adults. How many were children?

240 (borrow: 1, 1) − 70 = 170

170 children

G Last month our teacher read 90 pages of a mystery story to our class. This month he has read 35 more pages of the story. How many pages has the teacher read?

90 + 35 = 125

125 pages

8

	Borrow marks	Problem	Answer
A	5 10 9 1	6100 − 538	5562
B	2 9 1	300 − 94	206
C	6 9 14 1	7052 − 1378	5674
D		945 + 56	1001
E	2 10 9 1	3100 − 259	2841
F	2 1	4300 − 4240	60
G	7 11 9 1	8206 − 7949	257
H	7 1	800 − 460	340
I		78 + 40	118
J	2 18 1	3928 − 2998	930

Lesson 46

Facts + Problems + Bonus = TOTAL

1

11 − 2 = 9 7 − 2 = 5 9 − 2 = 7 5 − 2 = 3 4 − 2 = 2 10 − 2 = 8 6 − 2 = 4 8 − 2 = 6

10 − 2 = 8 3 − 2 = 1 7 − 2 = 5 6 − 2 = 4 11 − 2 = 9 8 − 2 = 6 5 − 2 = 3 9 − 2 = 7

2

15 − 6 = 9 15 − 9 = 6 15 − 7 = 8 15 − 8 = 7 15 − 6 = 9 15 − 10 = 5 15 − 7 = 8 15 − 8 = 7

15 − 7 = 8 15 − 8 = 7 15 − 6 = 9 15 − 9 = 6 15 − 8 = 7 15 − 10 = 5 15 − 6 = 9 15 − 7 = 8

3

13 − 5 = 8 11 − 9 = 2 13 − 4 = 9 12 − 9 = 3 13 − 8 = 5 10 − 4 = 6 13 − 9 = 4 13 − 6 = 7

9 − 4 = 5 16 − 7 = 9 14 − 9 = 5 8 − 3 = 5 9 − 3 = 6 13 − 5 = 8 16 − 7 = 9 10 − 2 = 8

9 − 6 = 3 12 − 4 = 8 13 − 4 = 9 9 − 7 = 2 13 − 9 = 4 10 − 8 = 2 13 − 6 = 7 16 − 9 = 7

9 − 4 = 5 15 − 9 = 6 10 − 7 = 3 8 − 5 = 3 17 − 9 = 8 13 − 4 = 9 16 − 7 = 9 12 − 3 = 9

4

A 800 − 1 = 799 B 500 − 1 = 499 C 100 − 1 = 99

D 600 − 1 = 599 E 200 − 1 = 199 F 900 − 1 = 899

Lesson 46 (continued)

5

A There were 36 nuts in a jar. Some more nuts were put into the jar. Now there are 52 nuts in the jar. How many more nuts were put into the jar?

52 (borrow: 4, 1) − 36 = 16

16 nuts

B Gail's science class was studying insects. In one week Gail collected 14 insects. The next week she collected another 20 insects. How many did she have then?

14 + 20 = 34

34 insects

C Gus had learned to play 16 songs on the violin. Then he learned how to play some new songs. Now Gus can play 27 songs. How many new songs did he learn?

27 − 16 = 11

11 songs

D A school had a spring sports contest. 45 students arrived early. Then later 160 more students arrived. How many students came to the contest?

45 + 160 = 205

205 students

E Janet raises rabbits. In March, Janet owned 34 rabbits. In May, she owned 70. How many new rabbits did she get from March to May?

70 (borrow: 6, 1) − 34 = 36

36 rabbits

6

A A waiter served 64 hot drinks to people in a restaurant. He served 90 drinks in all. How many cold drinks did the waiter serve?

90 (borrow: 8, 1) − 64 = 26

26 cold drinks

Part 6 continues on the next page.

B During a sale, a men's shop sold 90 leather jackets and 42 wool jackets. How many jackets did the shop sell?

[132] jackets

90 + 42 = 132

C On a ranch there are 90 full-grown sheep and 42 lambs. How many animals are on the ranch?

[132] animals

90 + 42 = 132

D There were 90 adult monkeys in a zoo. The zoo had 140 monkeys in all. How many baby monkeys did the zoo have?

[50] baby monkeys

140 − 90 = 50

E 530 people work in a factory. On Mondays 80 people leave to go to classes. How many people are left at work?

[450] people

(4,) 530 − 80 = 450

F Janet collects coins. On Monday she owned 34 coins. On Friday she owned 70 coins. How many coins did she get from Monday to Friday?

[36] coins

(6,) 70 − 34 = 36

G We grew roses and daisies in our garden. We picked 3 daisies. We picked 20 flowers in all. How many roses did we pick?

[17] roses

(1,) 20 − 3 = 17

H In an office there were 940 workers. 80 more workers are hired. How many people work in the office now?

[1020] people

940 + 80 = 1020

7

	A	B	C	D	E
borrow	79,	6'3,		8,	3'09,
top	800	7400	3284	900	4106
bottom	− 42	− 6990	+ 820	− 80	− 1328
answer	**758**	**410**	**4104**	**820**	**2778**

	F	G	H	I	J
borrow	4'29,	7'29,	09,	89,	
top	5300	8300	3104	7902	3842
bottom	− 869	− 974	− 28	− 5896	+ 1060
answer	**4431**	**7326**	**3076**	**2006**	**4902**

	K	L	M	N	O
borrow		89'0,	4,	3,	8,
top	5600	9012	5340	8040	9012
bottom	+ 699	− 8996	− 4930	− 1036	− 8612
answer	**6299**	**16**	**410**	**7004**	**400**

Lesson 47

Facts + Problems + Bonus = TOTAL

1

11	11	11	11	11	11	11	11
− 2	− 5	− 3	− 4	− 3	− 5	− 2	− 5
9	**6**	**8**	**7**	**8**	**6**	**9**	**6**

2

A 15 { 9, [6] — 15 − 9 = 6; 15 − 6 = 9

B 15 { 8, [7] — 15 − 8 = 7; 15 − 7 = 8

3

15	15	15	15	15	15	15	15
− 6	− 10	− 8	− 6	− 9	− 7	− 6	− 9
9	**5**	**7**	**9**	**6**	**8**	**9**	**6**
15	15	15	15	15	15	15	15
− 8	− 6	− 10	− 7	− 9	− 6	− 7	− 8
7	**9**	**5**	**8**	**6**	**9**	**8**	**7**

4

8	13	13	9	16	13	9	13
− 2	− 4	− 9	− 2	− 7	− 5	− 4	− 6
6	**9**	**4**	**7**	**9**	**8**	**5**	**7**
10	13	9	13	12	6	13	9
− 4	− 7	− 5	− 8	− 9	− 2	− 6	− 6
6	**6**	**4**	**5**	**3**	**4**	**7**	**3**
13	9	11	13	13	13	15	9
− 5	− 5	− 9	− 9	− 4	− 8	− 9	− 7
8	**4**	**2**	**4**	**9**	**5**	**6**	**2**
16	14	13	10	17	10	13	14
− 7	− 6	− 7	− 6	− 9	− 8	− 5	− 8
9	**8**	**6**	**4**	**8**	**2**	**8**	**6**

5

A 800 − 1 = **799** **B** 400 − 1 = **399** **C** 100 − 1 = **99**

D 600 − 1 = **599** **E** 900 − 1 = **899** **F** 300 − 1 = **299**

6

A In a park there is a goldfish pond. 48 goldfish were in the pond. This spring 70 more goldfish were put in the pond. How many goldfish are in the pond now?

[118] goldfish

48 + 70 = 118

B There was a bad storm last night. 64 trees were smashed in our town. Some more trees were smashed in the next town. 80 trees in all were smashed. How many trees were smashed in the next town?

[16] trees

(7,) 80 − 64 = 16

C By six o'clock a restaurant had served 14 pizzas. Then it served some more pizzas. By ten o'clock, 22 pizzas had been served. How many pizzas were served from six o'clock until ten o'clock?

[8] pizzas

(1,) 22 − 14 = 8

D At the beginning of the year, 14 swimmers were on the team. Later some more swimmers joined the team. At the end of the year, there were 22 swimmers on the team. How many swimmers joined the team during the year?

[8] swimmers

(1,) 22 − 14 = 8

E A country fair was held in our town yesterday. 1468 people were there very early. 1680 more people came later. How many people were at the fair then?

[3148] people

1468 + 1680 = 3148

Part 6 continues on the next page.

F Yoko was helping her father build a brick wall. By the beginning of May, they had used 384 bricks. They used some more bricks. By the end of May 830 bricks had been used in building the wall. How many bricks did Yoko and her father use during the month of May?

830 − 384 = 446

446 bricks

7

A There is a total of 250 students who go to Grant School. 208 students are present today. How many students are absent?

250 − 208 = 42

42 students

B A train was traveling from Toronto to Montreal. There were 145 adults on the train. There were 220 passengers altogether. How many children were on the train?

220 − 145 = 75

75 children

C There were 100 clocks in a clock shop. The shopowner bought 84 more clocks. How many clocks are in the shop now?

100 + 84 = 184

184 clocks

D 145 people work in the car factory. 25 more people have been hired to work on Monday. How many people in all will be working on Monday?

145 + 25 = 170

170 people

Part 7 continues on the next page.

E Mrs. Franks and Mrs. Schultz sold popcorn. One morning Mrs. Franks sold 20 bags of popcorn. Mrs. Schultz sold 40 bags. How many bags of popcorn did they sell?

20 + 40 = 60

60 bags

F Jenny has 408 Canadian stamps. She traded 98 stamps for some coins. How many Canadian stamps does Jenny have now?

408 − 98 = 310

310 stamps

G A repair shop fixed 48 watches in the morning. It repaired 90 watches in all that day. How many watches did it repair during the afternoon?

90 − 48 = 42

42 watches

H A market sold 2350 apples. It sold 2150 oranges. How many pieces of fruit did the market sell?

2350 + 2150 = 4500

4500 pieces

8

A	B	C	D	E
5042 − 4980 = 62	8700 − 7984 = 716	3510 − 3410 = 100	500 − 80 = 420	7406 − 970 = 6436

F	G	H	I	J
908 + 38 = 946	600 − 43 = 557	8100 − 299 = 7801	7058 − 950 = 6108	426 − 99 = 327

Lesson 48

Test + Facts + Problems + Bonus = TOTAL

1

11 − 2 = 9 | 11 − 5 = 6 | 11 − 3 = 8 | 11 − 4 = 7 | 11 − 2 = 9 | 11 − 4 = 7 | 11 − 3 = 8 | 11 − 5 = 6

2

A 14: 9, 5 — 14 − 9 = 5; 14 − 5 = 9

B 17: 9, 8 — 17 − 9 = 8; 17 − 8 = 9

C 16: 9, 7 — 16 − 9 = 7; 16 − 7 = 9

D 12: 9, 3 — 12 − 9 = 3; 12 − 3 = 9

3

15 − 6 = 9 | 15 − 10 = 5 | 15 − 7 = 8 | 15 − 9 = 6 | 15 − 8 = 7 | 15 − 6 = 9 | 15 − 9 = 6 | 15 − 7 = 8

15 − 10 = 5 | 15 − 7 = 8 | 15 − 6 = 9 | 15 − 9 = 6 | 15 − 8 = 7 | 15 − 7 = 8 | 15 − 6 = 9 | 15 − 9 = 6

4

8 − 2 = 6 | 13 − 4 = 9 | 7 − 2 = 5 | 9 − 5 = 4 | 13 − 5 = 8 | 8 − 5 = 3 | 9 − 2 = 7 | 13 − 7 = 6

13 − 8 = 5 | 10 − 2 = 8 | 9 − 5 = 4 | 13 − 6 = 7 | 12 − 9 = 3 | 9 − 4 = 5 | 13 − 7 = 6 | 14 − 9 = 5

13 − 9 = 4 | 6 − 2 = 4 | 9 − 4 = 5 | 13 − 5 = 8 | 11 − 2 = 9 | 9 − 6 = 3 | 13 − 4 = 9 | 9 − 7 = 2

Part 4 continues on the next page.

13 − 8 = 5 | 16 − 9 = 7 | 16 − 7 = 9 | 9 − 3 = 6 | 13 − 4 = 9 | 15 − 9 = 6 | 13 − 5 = 8 | 17 − 9 = 8

5

A	B	C	D
6000 − 4184 = 1816	9003 − 1248 = 7755	1004 − 368 = 636	4000 − 82 = 3918

6

A Gretchen had a cup collection. When the year began, she had 164 cups. At the end of the year, she had 270 cups. How many cups did she get during the year?

270 − 164 = 106

106 cups

B James had typed 114 pages of a report. Then he typed some more pages. Now he has typed 142 pages. How many more pages did James type?

142 − 114 = 28

28 pages

C My uncle could play 118 songs on the piano. This year he learned to play 64 more songs. How many songs can my uncle play now?

118 + 64 = 182

182 songs

D There were 143 ducks swimming in a lake. Some more ducks flew down to the lake. Then there were 160 ducks on the lake. How many ducks flew down to the lake?

160 − 143 = 17

17 ducks

Part 6 continues on the next page.

Lesson 48 (continued)

E Mr. Simmons is a secretary in a large office. On Monday he mailed 34 letters. On Tuesday he mailed 60 letters. How many letters did Mr. Simmons mail on Monday and Tuesday?

34
+ 60
94

[94] letters

7

A Our team lost 42 games. It played 70 games in all. How many games did our team win?

6,
70
− 42
28

[28] games

B A farmer had 136 bunches of carrots. He pulled up 16 more bunches of carrots. How many bunches of carrots did the farmer have then?

136
+ 16
152

[152] bunches

C 140 paintings were on sale at an art show. Some of the paintings were sold. At the end of the show there were 85 paintings left. How many were sold?

0'3,
140
− 85
55

[55] paintings

D Mr. Galanos has entered many kinds of contests. He has won prizes in 36 contests. In 42 other contests, he didn't win any prizes. How many contests in all has Mr. Galanos entered?

36
+ 42
78

[78] contests

E A pet shop sells dog collars and cat collars. It has 64 dog collars. The shop has 80 collars in all. How many cat collars does the shop have?

7,
80
− 64
16

[16] cat collars

Part 7 continues on the next page.

Lesson 48 (continued)

F Margo and Helene are building a table. Margo has used 20 nails so far in building the table. Helene has used 15 nails. How many nails have the girls used?

20
+ 15
35

[35] nails

G 406 people were at the dance. 9 people went home early. How many people were still at the dance?

39,
406
− 9
397

[397] people

8

	A	B	C	D	E
	2'4,	49,	2'49,	3,	3'2,
	4380	500	3506	400	4308
	− 4273	− 86	− 719	− 60	− 3530
	77	414	2787	340	778

	F	G	H	I	J
	3'29,	6'09,	7,	19,	2'0,
	4300	7100	800	3200	3100
	− 926	− 824	− 60	− 3184	− 1440
	3374	6276	740	16	1660

Lesson 49

Facts + Problems + Bonus = TOTAL

1

11	11	11	11	11	11	11	11
− 4	− 3	− 5	− 2	− 4	− 2	− 4	− 5
7	8	6	9	7	9	7	6

2

A 14 { 9 14 − 9 = 5 ; [5] 14 − 5 = 9 }

B 17 { 9 17 − 9 = 8 ; [8] 17 − 8 = 9 }

C 16 { 9 16 − 9 = 7 ; [7] 16 − 7 = 9 }

D 13 { 9 13 − 9 = 4 ; [4] 13 − 4 = 9 }

3

14	15	15	15	14	16	15	13
− 5	− 7	− 9	− 6	− 5	− 9	− 8	− 4
9	8	6	9	9	7	7	9
14	17	16	13	15	14	15	16
− 9	− 9	− 7	− 4	− 7	− 5	− 9	− 7
5	8	9	9	8	9	6	9

4

9	15	13	11	15	13	15	13
− 2	− 6	− 4	− 2	− 8	− 6	− 7	− 8
7	9	9	9	7	7	8	5
13	15	13	7	13	15	17	13
− 6	− 9	− 5	− 2	− 9	− 6	− 9	− 9
7	6	8	5	4	9	8	4
14	9	15	15	9	11	10	16
− 9	− 7	− 7	− 8	− 6	− 9	− 2	− 9
5	2	8	7	3	2	8	7

Part 4 continues on the next page.

Lesson 49 (continued)

12	13	15	13	15	13	13	9
− 9	− 5	− 6	− 4	− 8	− 7	− 6	− 4
3	8	9	9	7	6	7	5

5

	A	B	C	D	E
	499,	49,	4,	799,	79,
	5000	5000	5000	8000	8000
	− 3	− 30	− 300	− 4	− 40
	4997	4970	4700	7996	7960

	F	G	H	I	J
	299,	2,	99,	9,	
	3000	3000	1000	1000	1000
	− 8	− 800	− 7	− 20	− 400
	2992	2200	993	980	600

6

A There were 48 horses on the ranch. Some horses were born. Now there are 66 horses. How many horses were born?

5,
66
− 48
18

[18] horses

B This year a shop sold 1463 gray suits. The shop also sold 3652 blue suits. How many suits did the shop sell?

1463
+ 3652
5115

[5115] suits

C A peach grower sent 400 boxes of peaches to a market. 25 boxes were spoiled when they arrived. How many boxes were all right?

39,
400
− 25
375

[375] boxes

D Mrs. Weston collects plates. When she started, she had 134 plates. At the end of a year Mrs. Weston had 200 plates. How many plates did she get during the year?

19,
200
− 134
66

[66] plates

Part 6 continues on the next page.

Lesson 49 (continued)

E Mrs. Carr usually exercises 142 hours every month. Last month she missed some hours. She only exercised for 126 hours. How many hours of exercise did Mrs. Carr miss last month?

142 − 126 = 16

16 hours

F Bryan invited 13 children and 20 adults to his mother's birthday party. How many people were invited to the party?

13 + 20 = 33

33 people

G An automobile racing club had 84 members. 14 more members joined the club. How many people belong to the club now?

84 + 14 = 98

98 people

H 143 women went to the town meeting. 280 people in all attended the meeting. How many men attended the meeting?

280 − 143 = 137

137 men

I Mr. Teller owns a school that trains dogs. At the beginning of June 369 dogs had been trained. By the end of June 435 dogs had been trained. How many dogs were trained in June?

435 − 369 = 66

66 dogs

7

A	B	C	D	E
6400 − 986 = 5414	7100 − 280 = 6820	5120 − 4370 = 750	6012 − 4999 = 1013	4106 − 9 = 4097

F	G	H	I	J
5600 + 699 = 6299	9012 − 8996 = 16	5340 − 4930 = 410	8040 − 1036 = 7004	9012 − 8612 = 400

Lesson 50

Test + Facts + Problems + Bonus = TOTAL

1

11 − 2 = 9	11 − 4 = 7	11 − 3 = 8	11 − 5 = 6	11 − 3 = 8	11 − 2 = 9	11 − 5 = 6	11 − 4 = 7

2

A 14 {9, [5]}: 14 − 9 = 5; 14 − 5 = 9

B 16 {9, [7]}: 16 − 9 = 7; 16 − 7 = 9

3

14 − 9 = 5	16 − 7 = 9	14 − 5 = 9	14 − 8 = 6	16 − 7 = 9	16 − 8 = 8	13 − 8 = 5	14 − 5 = 9
13 − 4 = 9	16 − 7 = 9	13 − 9 = 4	14 − 6 = 8	13 − 4 = 9	16 − 8 = 8	13 − 9 = 4	14 − 5 = 9

4

15 − 6 = 9	13 − 8 = 5	8 − 2 = 6	13 − 6 = 7	15 − 7 = 8	9 − 2 = 7	11 − 2 = 9	15 − 8 = 7
13 − 7 = 6	9 − 2 = 7	13 − 5 = 8	16 − 8 = 8	15 − 8 = 7	7 − 2 = 5	13 − 7 = 6	15 − 9 = 6
10 − 4 = 6	6 − 2 = 4	15 − 8 = 7	13 − 4 = 9	16 − 8 = 8	15 − 7 = 8	13 − 6 = 7	5 − 4 = 1
13 − 8 = 5	10 − 3 = 7	15 − 6 = 9	12 − 5 = 7	4 − 2 = 2	13 − 5 = 8	12 − 8 = 4	13 − 4 = 9

Lesson 50 (continued)

5

A 90 children went to a play. 46 of the children enjoyed the play. The rest did not. How many of the children did not enjoy the play?

90 − 46 = 44

B We bought some eggs. When we got home, 36 eggs were broken. 24 eggs were not broken. How many eggs in all did we buy?

36 + 24 = 60

C There are 62 sharks. 45 of the sharks are hungry. How many of the sharks are not hungry?

62 − 45 = 17

D 80 people were watching a bulldozer knock down a building. 69 people left to return to work. How many people stayed to watch?

80 − 69 = 11

E Linda Arrowhead's father is a baker. One Saturday Linda helped her father make pies for some restaurants. They made 60 apple pies and 45 pies that were not apple. How many pies in all did Linda and her father make?

60 + 45 = 105

6

A	B	C	D	E
3000 − 8 = 2992	3000 − 800 = 2200	3000 − 80 = 2920	1000 − 20 = 980	1000 − 2 = 998

F	G	H	I	J
4000 − 245 = 3755	4000 − 2450 = 1550	8600 − 342 = 8258	8000 − 300 = 7700	6000 − 4128 = 1872

Lesson 50 (continued)

7

A 136 people had paid for swimming lessons. Then some more people paid for swimming lessons. Now 180 people have paid for lessons. How many more people paid for swimming lessons?

180 − 136 = 44

44 people

B A greeting card shop sold 43 cards in the morning. It sold 50 cards in the afternoon. How many cards in all did it sell?

43 + 50 = 93

93 cards

C There were 3170 cars in the parking garage. There were 990 dirty cars. How many clean cars were there?

3170 − 990 = 2180

2180 clean cars

D When we first moved to Morris, 2683 persons lived here. Now 3780 people live in Morris. How many people have moved to Morris?

3780 − 2683 = 1097

1097 people

E Carmen ran for 37 hours this week. Elena ran for 40 hours this week. How many hours in all did the girls run this week?

37 + 40 = 77

77 hours

F Reggie had 70 old English coins. He gave some of them to his brother. Now Reggie has 48 old English coins. How many coins did he give to his brother?

70 − 48 = 22

22 coins

Part 7 continues on the next page.

Lesson 50 (continued)

G Mr. Yamada visits a used-book shop every week. Mr. Yamada had 36 books. Last week he bought some more books at the used-book shop. Now he has 58 books. How many books did he buy last week?

58 − 36 = 22

22 books

H An office building has 2365 clean windows. The window washers have to wash 90 dirty windows. How many windows in all does the building have?

2365 + 90 = 2455

2455 windows

8

	A	B	C	D	E
	3104	4106	3460	5360	4300
−	78	256	1990	5298	3996
	3026	**3850**	**1470**	**62**	**304**

	F	G	H	I	J
	5300	8300	3104	7902	7400
−	869	974	28	5896	6990
	4431	**7326**	**3076**	**2006**	**410**

Lesson 51

Facts + Problems + Bonus = TOTAL

1

11	11	11	11	11	11	11	11
− 2	− 3	− 5	− 4	− 3	− 5	− 4	− 2
9	**8**	**6**	**7**	**8**	**6**	**7**	**9**

2

A 17 { 9 — 17 − 9 = 8; **8** — 17 − 8 = 9

B 14 { 9 — 14 − 9 = 5; **5** — 14 − 5 = 9

3

17	14	17	13	14	18	16	12
− 9	− 6	− 8	− 6	− 5	− 9	− 7	− 9
8	**8**	**9**	**7**	**9**	**9**	**9**	**3**
14	11	14	13	15	17	15	16
− 5	− 9	− 8	− 4	− 9	− 8	− 9	− 7
9	**2**	**6**	**9**	**6**	**9**	**6**	**9**

4

15	11	13	10	15	13	8	15
− 6	− 2	− 8	− 2	− 7	− 6	− 2	− 8
9	**9**	**5**	**8**	**8**	**7**	**6**	**7**
14	15	13	9	13	11	15	9
− 9	− 7	− 9	− 2	− 6	− 9	− 8	− 2
5	**8**	**4**	**7**	**7**	**2**	**7**	**7**
13	6	13	15	12	13	5	13
− 5	− 2	− 8	− 7	− 9	− 7	− 2	− 5
8	**4**	**5**	**8**	**3**	**6**	**3**	**8**
13	9	13	16	15	9	13	9
− 6	− 4	− 8	− 9	− 6	− 3	− 5	− 6
7	**5**	**5**	**7**	**9**	**6**	**8**	**3**

Lesson 51 (continued)

5

A My sister made a beaded belt. She used 140 large beads to make the belt. 90 of the beads are red. How many of the beads are not red?

140 − 90 = **50**

B There's a new statue in our city. Lots of people saw the statue. 80 people liked the statue. 57 people did not like the statue. How many people saw the statue?

80 + 57 = **137**

C There are 140 people on the beach. 95 of them are wearing hats. How many of the people are not wearing hats?

140 − 95 = **45**

D There were 86 earthquakes last year. 29 of the earthquakes were in Europe. How many earthquakes were not in Europe?

86 − 29 = **57**

E We made sandwiches for our club's picnic. The people who came ate 95 ham sandwiches and 120 chicken sandwiches. How many sandwiches were eaten at the picnic?

95 + 120 = **215**

6

	A	B	C	D	E
	8000	8000	4000	4000	1000
−	3	30	352	3500	80
	7997	**7970**	**3648**	**500**	**920**

	F	G	H	I	J
	1000	7000	5000	3000	1000
−	8	1430	1008	125	946
	992	**5570**	**3992**	**2875**	**54**

Lesson 51 (continued)

7

A Our family is very large. We have 14 first cousins and 20 second cousins. How many cousins are there in our family?

14 + 20 = 34

34 cousins

B The racing team won 47 car races. They entered 52 car races. How many races did the team lose?

52 − 47 = 5

5 races

C At our club's party there were 42 children and 70 adults. How many people came to the party?

42 + 70 = 112

112 people

D Linda is collecting rocks. This morning she had 136 rocks. Tonight she has 142 rocks. How many rocks did she get during the day?

142 − 136 = 6

6 rocks

E Benny bought 420 file folders. 254 of the folders have been used. How many are new?

420 − 254 = 166

166 folders

F Sergio worked 36 crossword puzzles. Adolfo worked 40 crossword puzzles. How many puzzles did the boys work in all?

36 + 40 = 76

76 puzzles

Part 7 continues on the next page.

G Our stamp club had 126 members last year. Some more people joined the club. Now there are 142 members in the club. How many people joined the club?

142 − 126 = 16

16 people

H A friend of mine has read 462 mystery books and some sports books. She has read 500 books in all. How many sports books has she read?

500 − 462 = 38

38 sports books

I Don counted 24 sailboats on the lake yesterday. Today Don counted 30 sailboats on the lake. How many boats did Don see on the lake?

24 + 30 = 54

54 boats

8

A	B	C	D	E
5100 − 86 = 5014	3200 − 2164 = 1036	8100 − 364 = 7736	4600 − 3920 = 680	5400 − 1923 = 3477

Facts + Problems + Bonus = TOTAL

1

17 − 8 = 9	14 − 7 = 7	17 − 9 = 8	13 − 9 = 4	17 − 8 = 9	16 − 8 = 8	15 − 9 = 6	13 − 4 = 9
15 − 6 = 9	12 − 9 = 3	14 − 9 = 5	17 − 8 = 9	13 − 7 = 6	14 − 9 = 5	16 − 7 = 9	14 − 5 = 9

2

A 11 { 4, **7** — 11 − 4 = 7; 11 − 7 = 4

B 11 { 2, **9** — 11 − 2 = 9; 11 − 9 = 2

C 11 { 5, **6** — 11 − 5 = 6; 11 − 6 = 5

D 11 { 3, **8** — 11 − 3 = 8; 11 − 8 = 3

3

11 − 6 = 5	11 − 8 = 3	11 − 5 = 6	11 − 7 = 4	11 − 8 = 3	11 − 6 = 5	11 − 4 = 7	11 − 7 = 4
11 − 7 = 4	11 − 9 = 2	11 − 6 = 5	11 − 8 = 3	11 − 7 = 4	11 − 2 = 9	11 − 6 = 5	11 − 3 = 8

4

17 − 8 = 9	13 − 5 = 8	8 − 2 = 6	15 − 6 = 9	13 − 6 = 7	11 − 2 = 9	6 − 2 = 4	15 − 8 = 7
16 − 9 = 7	15 − 7 = 8	11 − 9 = 2	10 − 4 = 6	17 − 9 = 8	12 − 8 = 4	14 − 9 = 5	13 − 5 = 8

Part 4 continues on the next page.

15 − 9 = 6	9 − 2 = 7	12 − 3 = 9	13 − 7 = 6	13 − 6 = 7	15 − 7 = 8	13 − 9 = 4	5 − 2 = 3
17 − 9 = 8	13 − 8 = 5	12 − 9 = 3	7 − 2 = 5	15 − 9 = 6	13 − 6 = 7	15 − 8 = 7	16 − 9 = 7

5

A Melanie did 40 subtraction problems. 17 of them are wrong. How many of Melanie's problems are not wrong?

40 − 17 = 23

23 problems

B 137 people brought food to the picnic. 140 people did not bring food. How many people in all went to the picnic?

137 + 140 = 277

277 people

C 300 runners entered a race. 27 runners left the race. How many runners did not leave the race?

300 − 27 = 273

273 runners

D There were 68 people in front of city hall. 14 of them were police officers. How many people were not police officers?

68 − 14 = 54

54 people

E On Monday 48 people in a large office came to work on time. 7 people did not come to work on time. How many people in all worked in the office on Monday?

48 + 7 = 55

55 people

6

A	B	C	D	E
8000 − 4 = 7996	9000 − 30 = 8970	6000 − 235 = 5765	1000 − 800 = 200	1000 − 927 = 73

F	G	H	I	J
4000 − 120 = 3880	6000 − 305 = 5695	2000 − 125 = 1875	4000 − 300 = 3700	1000 − 120 = 880

7

A A bicycle factory made 4380 bicycles. Then they made another 1560 bicycles. How many bicycles did the factory make?

4380 + 1560 = 5940

5940 bicycles

B 140 people went fishing at a large lake. 65 of them caught fish. How many people did not catch any fish?

140 − 65 = 75

75 people

C There were 199 girls inside the school. Some more girls entered the school. Now there are 215 girls inside the school. How many more girls entered the school?

215 − 199 = 16

16 girls

D At a band concert there are 1824 men and 2400 women. How many people are at the concert?

1824 + 2400 = 4224

4224 people

Part 7 continues on the next page.

E Our team won 34 games. But the team also lost 30 games. How many games did the team play?

34 + 30 = 64

64 games

F An apple picker picked 147 apples. Then he picked some more apples. Now he has 180 apples. How many more apples did the apple picker pick?

180 − 147 = 33

33 apples

G 1460 girls were in a parade. 2320 children were in the parade. How many boys were in the parade?

2320 − 1460 = 860

860 boys

H On a cattle ranch there are 458 cows. There are also 280 bulls on the ranch. How many animals are there on the ranch?

458 + 280 = 738

738 animals

8

A. 6104 − 29 = 6075
B. 8107 − 352 = 7755
C. 210 − 74 = 136
D. 846 + 352 = 1198
E. 9062 − 8980 = 82

Lesson 53

Test + Facts + Problems + Bonus = TOTAL

1

11 − 8 = 3	11 − 2 = 9	11 − 5 = 6	11 − 3 = 8	11 − 4 = 7	11 − 7 = 4	11 − 9 = 2	11 − 6 = 5
11 − 5 = 6	11 − 7 = 4	11 − 6 = 5	11 − 4 = 7	11 − 8 = 3	11 − 3 = 8	11 − 9 = 2	11 − 7 = 4

2

A. 16 { 9, **7** } 16 − 9 = 7; 16 − 7 = 9
B. 13 { 9, **4** } 13 − 9 = 4; 13 − 4 = 9
C. 15 { 9, **6** } 15 − 9 = 6; 15 − 6 = 9
D. 17 { 9, **8** } 17 − 9 = 8; 17 − 8 = 9

3

13 − 4 = 9	17 − 9 = 8	15 − 8 = 7	15 − 6 = 9	13 − 8 = 5	16 − 7 = 9	12 − 9 = 3	14 − 5 = 9
15 − 9 = 6	14 − 8 = 6	16 − 7 = 9	13 − 9 = 4	18 − 9 = 9	14 − 5 = 9	17 − 9 = 8	13 − 4 = 9

4

13 − 9 = 4	15 − 7 = 8	13 − 6 = 7	9 − 4 = 5	15 − 8 = 7	13 − 7 = 6	15 − 6 = 9	15 − 7 = 8
10 − 4 = 6	15 − 7 = 8	15 − 8 = 7	13 − 7 = 6	13 − 5 = 8	15 − 8 = 7	13 − 6 = 7	9 − 5 = 4

Part 4 continues on the next page.

13 − 5 = 8	9 − 6 = 3	15 − 7 = 8	8 − 2 = 6	13 − 8 = 5	9 − 6 = 3	8 − 2 = 6	15 − 8 = 7
10 − 4 = 6	15 − 9 = 6	16 − 9 = 7	8 − 3 = 5	10 − 7 = 3	13 − 9 = 4	10 − 2 = 8	8 − 5 = 3

5

A There were 140 people on the train. 97 of them were reading a newspaper. How many people were not reading a newspaper?

140 − 97 = 43

43 people

B A reporter talked to some people at an art show. 48 people said they liked the art show. 72 people said they did not like the art show. How many people did the reporter talk to?

48 + 72 = 120

120 people

C Mrs. Kiyo planted 18 flower seeds. She also planted 40 seeds that were not flower seeds. How many seeds did Mrs. Kiyo plant?

18 + 40 = 58

58 seeds

D Natasha counted 94 cars. 87 of them were new cars. How many were not new cars?

94 − 87 = 7

7 cars

E There are 150 students. 79 students entered the spelling contest. How many students did not enter the spelling contest?

150 − 79 = 71

71 students

6

A. 6000 − 300 = 5700
B. 4000 − 1284 = 2716
C. 3000 − 18 = 2982
D. 1000 − 20 = 980
E. 6000 − 150 = 5850
F. 1000 − 246 = 754
G. 3000 − 400 = 2600
H. 8000 − 3460 = 4540
I. 6000 − 145 = 5855
J. 9000 − 8820 = 180

7

A Felipe collected bottles. He had 564 bottles. He found some more bottles. Now he has 610 bottles. How many more bottles did he find?

610 − 564 = 46

46 bottles

B Our principal, Miss DiBono, gave prizes to 14 boys. She gave prizes to 28 girls. How many children got prizes?

14 + 28 = 42

42 children

C Mr. Collins is a bookseller. He sold 420 sports books. He also sold 680 mystery books. How many books did he sell?

420 + 680 = 1100

1100 books

D A zoo had 385 birds that sang. The zoo also had 37 other birds that talked. How many birds in all did the zoo have?

385 + 37 = 422

422 birds

Part 7 continues on the next page.

E The store bought 1640 white boxes. It also bought some brown boxes. It bought 2100 boxes in all. How many brown boxes did the store buy?

	2100
−	1640
	460

460 brown boxes

F 143 shops are open on Sunday. 583 shops are not open. How many shops are there in all?

	143
+	583
	726

726 shops

G A flower shop had a big sale. It sold 436 flowers on the first day. By the end of the second day it had sold 544 flowers in all. How many flowers were sold on the second day?

	544
−	436
	108

108 flowers

H An office building is being built near our school. Workers finished 94 floors of the building last month. This month they finished 19 more floors. Now the building is ready. How many floors does it have?

	94
+	19
	113

113 floors

Lesson 54

Test + Facts + Problems + Bonus = TOTAL

1

11	11	11	11	11	11	11	11
− 8	− 2	− 7	− 4	− 6	− 9	− 3	− 7
3	9	4	7	5	2	8	4
11	11	11	11	11	11	11	11
− 6	− 4	− 8	− 6	− 5	− 7	− 3	− 8
5	7	3	5	6	4	8	3

2

A 17 { 9: 17 − 9 = 8; [8]: 17 − 8 = 9

B 14 { 9: 14 − 9 = 5; [5]: 14 − 5 = 9

C 15 { 9: 15 − 9 = 6; [6]: 15 − 6 = 9

D 13 { 9: 13 − 9 = 4; [4]: 13 − 4 = 9

3

17	13	14	17	16	17	14	13
− 8	− 9	− 5	− 9	− 9	− 8	− 7	− 9
9	4	9	8	7	9	7	4
14	15	14	15	17	15	17	15
− 5	− 8	− 9	− 6	− 9	− 8	− 9	− 6
9	7	5	9	8	7	8	9

4

15	13	16	8	15	15	13	10
− 8	− 6	− 7	− 2	− 6	− 7	− 8	− 2
7	7	9	6	9	8	5	8
9	15	15	7	13	13	13	15
− 2	− 7	− 8	− 2	− 6	− 7	− 5	− 7
7	8	7	5	7	6	8	8

Part 4 continues on the next page.

13	6	13	13	9	15	13	12
− 8	− 4	− 7	− 4	− 4	− 8	− 6	− 3
5	2	6	9	5	7	7	9
13	15	9	15	13	10	10	15
− 8	− 7	− 4	− 8	− 5	− 6	− 3	− 8
5	8	5	7	8	4	7	7

5

A There were 62 people at the game. 45 people cheered for the blue team. How many people did not cheer for the blue team?

62 − 45 = 17

17 people

B 58 people on a boat were sick. 70 people were not sick. How many people were on the boat?

58 + 70 = 128

128 people

C 82 people are working on a ship. Today only 47 people came to work. How many people did not come to work?

82 − 47 = 35

35 people

D Some tree cutters cut 90 trees in a forest. They found bugs in 27 of the trees. How many trees did not have bugs?

90 − 27 = 63

63 trees

E We live in a large house with many windows. Today 25 of the windows are open. 15 windows are not open. How many windows does our house have?

25 + 15 = 40

40 windows

6

A A factory has made 5360 machines. The factory has sold 4890 machines. How many machines does the factory have left to sell?

5360 − 4890 = 470

470 machines

B When a large market opened, it gave away balloons. The market gave away 86 red balloons and 90 green balloons. How many balloons did the market give away?

86 + 90 = 176

176 balloons

C 98 people work in a big office. 18 of these people are men. How many of these people are women?

98 − 18 = 80

80 people

D There were 1485 liters of water in the swimming pool. More water was added to the pool. Now there are 2095 liters of water in the pool. How many liters of water were added?

2095 − 1485 = 610

610 liters

E On Sunday there were 148 fishing boats on the lake. There were also 473 sailboats on the lake. How many boats were on the lake?

148 + 473 = 621

621 boats

Part 6 continues on the next page.

F Mrs. Ito runs a newsstand in front of a big office building. In the morning she sold 138 newspapers. She sold some more papers in the afternoon. By the end of the day she had sold 150 newspapers. How many newspapers did she sell in the afternoon?

$150 - 138 = 12$

12 newspapers

G Mr. Campos is a guide in an art museum. Yesterday he was preparing to take 85 people through the museum. At the last moment 17 more people joined the group. How many people did Mr. Campos take through the museum?

$85 + 17 = 102$

102 people

7

A. 5000 − 1402 = **3598**	B. 3000 − 140 = **2860**	C. 1000 − 230 = **770**	D. 1000 − 156 = **844**	E. 6000 − 200 = **5800**
F. 9004 − 136 = **8868**	G. 3006 − 106 = **2900**	H. 4000 − 125 = **3875**	I. 1004 − 83 = **921**	J. 5003 − 3999 = **1004**
K. 4102 − 86 = **4016**	L. 5103 − 73 = **5030**	M. 8210 − 7946 = **264**	N. 920 − 814 = **106**	O. 7100 − 386 = **6714**

Facts + Problems + Bonus = TOTAL

1

11 − 7 = **4**	11 − 4 = **7**	11 − 5 = **6**	11 − 8 = **3**	11 − 3 = **8**	11 − 7 = **4**	11 − 6 = **5**	11 − 9 = **2**
11 − 4 = **7**	11 − 6 = **5**	11 − 2 = **9**	11 − 3 = **8**	11 − 8 = **3**	11 − 2 = **9**	11 − 5 = **6**	11 − 7 = **4**

2

A 13 { 9, **4** } — **13 − 9 = 4**; **13 − 4 = 9**

B 16 { 9, **7** } — **16 − 9 = 7**; **16 − 7 = 9**

C 17 { 9, **8** } — **17 − 9 = 8**; **17 − 8 = 9**

D 15 { 9, **6** } — **15 − 9 = 6**; **15 − 6 = 9**

3

12 − 9 = **3**	15 − 6 = **9**	13 − 9 = **4**	14 − 7 = **7**	17 − 8 = **9**	14 − 9 = **5**	15 − 6 = **9**	15 − 8 = **7**
16 − 7 = **9**	13 − 9 = **4**	13 − 5 = **8**	17 − 8 = **9**	16 − 9 = **7**	14 − 5 = **9**	13 − 8 = **5**	13 − 4 = **9**

4

13 − 5 = **8**	13 − 7 = **6**	9 − 4 = **5**	13 − 4 = **9**	15 − 7 = **8**	13 − 8 = **5**	15 − 8 = **7**	13 − 6 = **7**
9 − 5 = **4**	13 − 7 = **6**	9 − 2 = **7**	13 − 5 = **8**	16 − 9 = **7**	13 − 8 = **5**	9 − 6 = **3**	16 − 7 = **9**

Part 4 continues on the next page.

13 − 5 = **8**	15 − 8 = **7**	8 − 2 = **6**	10 − 6 = **4**	9 − 3 = **6**	13 − 8 = **5**	10 − 4 = **6**	15 − 7 = **8**
13 − 8 = **5**	15 − 7 = **8**	8 − 3 = **5**	10 − 6 = **4**	8 − 5 = **3**	10 − 4 = **6**	10 − 2 = **8**	14 − 6 = **8**

5

A Mrs. Savas is 42 years old. Miss Hark is 70 years old. How many years older is Miss Hark?

$70 - 42 = 28$

B After a few games of table tennis, Amos Silverheels had scored 26 points. His brother Mike had scored 40 points. How many points did Amos and Mike score in all?

$26 + 40 = 66$

C Gina weighs 48 kilograms. Sam weighs 29 kilograms. How many more kilograms does Gina weigh?

$48 - 29 = 19$

D Mel exercised for 48 minutes. Brad exercised for 42 minutes. How many minutes did Mel and Brad exercise in all?

$48 + 42 = 90$

E A building is 36 meters tall. A tree near the building is 48 meters tall. How many meters taller is the tree?

$48 - 36 = 12$

6

A Swimmers found 1500 coins underwater. 896 of the coins were gold. How many coins were not gold?

$1500 - 896 = 604$

604 coins

Part 6 continues on the next page.

B 1536 people were at the boat races. Then some more people arrived. Now there are 1600 people at the boat races. How many more people arrived?

$1600 - 1536 = 64$

64 people

C 1504 people took driving tests. 1486 people passed their driving tests. How many people did not pass their driving tests?

$1504 - 1486 = 18$

18 people

D Dan had 48 Canadian stamps. His cousin sent him 26 Canadian stamps. How many Canadian stamps does Dan have in all?

$48 + 26 = 74$

74 stamps

E Jennie jumped rope 140 times last week. Lisa jumped rope 362 times. How many times did Jennie and Lisa jump rope?

$140 + 362 = 502$

502 times

F Miss Benton answered 945 telephone calls on Monday. On Tuesday she answered some more telephone calls. Now she has answered 1200 telephone calls. How many telephone calls did Miss Benton answer on Tuesday?

$1200 - 945 = 255$

255 telephone calls

G 195 students are in classes at Oak School. 24 students are absent from classes. How many students in all go to Oak School?

$195 + 24 = 219$

219 students

Part 6 continues on the next page.

H The builders had 1530 nails. They used up most of the nails. Now they have 86 nails left. How many nails did they use?

$$1530 - 86 = 1444$$

1444 nails

I Last week we counted the cars that drove past our house. 3500 cars went by. 2850 cars followed the speed limit. How many cars did not follow the speed limit?

$$3500 - 2850 = 650$$

650 cars

J At one summer camp there were 1368 girls. There are 2100 children at the camp. How many of the children are boys?

$$2100 - 1368 = 732$$

732 children

7

A 5000 − 324 = **4676**	**B** 1000 − 230 = **770**	**C** 7204 − 264 = **6940**	**D** 3100 − 2134 = **966**	**E** 1000 − 364 = **636**
F 9200 − 235 = **8965**	**G** 6000 − 4200 = **1800**	**H** 8000 − 7940 = **60**	**I** 1000 − 429 = **571**	**J** 3005 − 2204 = **801**
K 6024 − 188 = **5836**	**L** 5100 − 480 = **4620**	**M** 7024 − 3614 = **3410**	**N** 8326 − 1996 = **6330**	**O** 5034 − 4628 = **406**

Facts + Problems + Bonus = TOTAL

1

11 − 6 = **5**	11 − 4 = **7**	11 − 8 = **3**	11 − 9 = **2**	11 − 6 = **5**	11 − 3 = **8**	11 − 8 = **3**	11 − 7 = **4**
11 − 5 = **6**	11 − 7 = **4**	11 − 3 = **8**	11 − 8 = **3**	11 − 6 = **5**	11 − 4 = **7**	11 − 9 = **2**	11 − 7 = **4**

2

A 15 { 9, 6 }: 15 − 9 = 6; 15 − 6 = 9

B 13 { 9, 4 }: 13 − 9 = 4; 13 − 4 = 9

C 17 { 9, 8 }: 17 − 9 = 8; 17 − 8 = 9

D 16 { 9, 7 }: 16 − 9 = 7; 16 − 7 = 9

3

12 − 6 = **6**	15 − 6 = **9**	13 − 9 = **4**	14 − 5 = **9**	16 − 7 = **9**	10 − 5 = **5**	15 − 6 = **9**	12 − 9 = **3**
17 − 8 = **9**	16 − 9 = **7**	16 − 10 = **6**	15 − 6 = **9**	13 − 9 = **4**	14 − 5 = **9**	12 − 9 = **3**	17 − 8 = **9**

4

13 − 5 = **8**	8 − 2 = **6**	9 − 3 = **6**	13 − 7 = **6**	9 − 5 = **4**	13 − 6 = **7**	15 − 8 = **7**	15 − 7 = **8**
13 − 7 = **6**	9 − 4 = **5**	13 − 5 = **8**	9 − 6 = **3**	16 − 7 = **9**	13 − 8 = **5**	9 − 3 = **6**	15 − 7 = **8**

Part 4 continues on the next page.

9 − 7 = **2**	15 − 8 = **7**	8 − 3 = **5**	13 − 6 = **7**	9 − 2 = **7**	8 − 5 = **3**	15 − 7 = **8**	14 − 5 = **9**
9 − 4 = **5**	13 − 7 = **6**	13 − 6 = **7**	15 − 8 = **7**	13 − 8 = **5**	10 − 6 = **4**	9 − 4 = **5**	6 − 2 = **4**

5

A Jana and Allison like to play marbles. Jana has 137 marbles. Allison has 170 marbles. How many more marbles does Allison have than Jana?

$$170 - 137 = 33$$

B My mother's new car weighs 1340 kilograms. My mother's old car weighed 1700 kilograms. How many kilograms do the cars weigh in all?

$$1340 + 1700 = 3040$$

C Mr. Mills is 45 years old. Mr. Young is 60 years old. How many years older is Mr. Young?

$$60 - 45 = 15$$

D Rafael and Matt worked as ticket sellers at a circus. Rafael sold 94 tickets and Matt sold 120 tickets. How many tickets did the boys sell?

$$94 + 120 = 214$$

E An apple tree is 594 centimeters tall. An oak tree is 730 centimeters tall. How many centimeters taller than the apple tree is the oak tree?

$$730 - 594 = 136$$

6

A The store sold 185 red pencils. It sold 480 black pencils. How many pencils did the store sell?

$$185 + 480 = 665$$

665 pencils

Part 6 continues on the next page.

B 154 cooks entered a cooking contest. 89 cooks won prizes. How many cooks did not win prizes?

$$154 - 89 = 65$$

65 cooks

C In a large market, shoppers were asked to taste a new kind of cake. 425 shoppers liked the cake. 128 shoppers did not like the cake. How many shoppers tasted the cake?

$$425 + 128 = 553$$

553 shoppers

D When a zoo got Emma the elephant, she weighed 2589 kilograms. Now Emma weighs 3420 kilograms. How many kilograms has Emma gained since she has been in the zoo?

$$3420 - 2589 = 831$$

831 kilograms

E There were 1365 boys at the school bicycle races. There were 2100 children at these races. How many of them were girls?

$$2100 - 1365 = 735$$

735 girls

F A farmer sent 548 potatoes to a market in the city. He also sent 379 onions. How many vegetables did the farmer send?

$$548 + 379 = 927$$

927 vegetables

G Ann likes to read mystery stories. In June she read 28 mystery stories. In July she read 17 mystery stories. How many mystery stories did she read?

$$28 + 17 = 45$$

45 mystery stories

Part 6 continues on the next page.

Lesson 56 (continued)

H Last year our family took 1444 photographs. 180 of them were black and white. How many of the photographs were not black and white?

1444 − 180 = 1264

1264 photographs

I There were 518 people at a dance. Some more people came to the dance. Then there were 590 people at the dance. How many more people came to the dance?

590 − 518 = 72

72 people

J 42 ducks landed on the pond on Sunday. On Tuesday 30 ducks landed on the pond. How many ducks landed on the pond?

42 + 30 = 72

72 ducks

K 520 people attended a concert. 34 people went home early. How many people stayed at the concert?

520 − 34 = 486

486 people

7

	A	B	C	D	E
	6003	4200	1800	1000	5006
	− 80	− 905	− 286	− 888	− 206
	5923	3295	1514	112	4800

	F	G	H	I	J
	3102	1000	4105	6000	5000
	−1996	− 200	−3999	−4990	−3500
	1106	800	106	1010	1500

Lesson 57

Facts + Problems + Bonus = TOTAL

1

15	15	15	15	15	15	15	15
− 8	− 6	− 9	− 7	−10	− 6	− 9	− 8
7	9	6	8	5	9	6	7
15	15	15	15	15	15	15	15
−10	− 7	− 9	− 8	− 6	− 7	− 8	− 6
5	8	6	7	9	8	7	9

2

A 13 { 9: 13 − 9 = 4; **4**: 13 − 4 = 9

B 17 { 9: 17 − 9 = 8; **8**: 17 − 8 = 9

C 14 { 9: 14 − 9 = 5; **5**: 14 − 5 = 9

D 11 { 9: 11 − 9 = 2; **2**: 11 − 2 = 9

3

14	16	15	13	17	14	17	15
− 5	− 8	− 9	− 4	− 9	− 7	− 8	− 8
9	8	6	9	8	7	9	7
14	11	13	17	16	12	15	17
− 5	− 9	− 9	− 8	− 8	− 9	− 7	− 8
9	2	4	9	8	3	8	9

4

11	13	15	11	13	17	13	11
− 7	− 9	− 6	− 8	− 5	− 8	− 7	− 6
4	4	9	3	8	9	6	5
13	11	15	13	11	13	13	11
− 5	− 9	− 7	− 4	− 6	− 6	− 8	− 7
8	2	8	9	5	7	5	4

Part 4 continues on the next page.

Lesson 57 (continued)

9	11	9	11	8	11	9	11
− 3	− 8	− 7	− 4	− 3	− 5	− 5	− 3
6	3	2	7	5	6	4	8
8	16	11	9	7	9	13	9
− 5	− 9	− 2	− 6	− 3	− 4	− 6	− 4
3	7	9	3	4	5	7	5

5

A We saw two horses pulling a wagon. Their owner said that one weighed 874 kilograms. The other weighed 920 kilograms. How many kilograms did the horses weigh in all?

874 + 920 = 1794

1794 kilograms

B Our city has 46 restaurants and 90 shops. How many more shops does the city have than restaurants?

90 − 46 = 44

44 shops

C The officers who keep track of animals in a national park put tags on 1482 deer and 1800 elk. How many tags were put on in all?

1482 + 1800 = 3282

3282 tags

D A young elephant weighs 1350 kilograms. Its mother weighs 2500 kilograms. How many more kilograms does the mother elephant weigh?

2500 − 1350 = 1150

1150 kilograms

E Mr. Welbes carved 62 wooden ducks. Mr. Kiyo carved 39 wooden ducks. How many more ducks did Mr. Welbes carve than Mr. Kiyo?

62 − 39 = 23

23 ducks

Lesson 57 (continued)

6

A Mr. Gallo planted 147 fruit trees on his ranch. 28 fruit trees did not grow. How many fruit trees did grow?

147 − 28 = 119

119 fruit trees

B Last week a baker sold 1350 blueberry muffins. This week the baker sold 1500 cherry muffins. How many muffins did the baker sell?

1350 + 1500 = 2850

2850 muffins

C Mr. Delgado teaches piano. He had 167 students. Then he got some more students. Now he has 210 students. How many more students did he get?

210 − 167 = 43

43 students

D 190 children came to school today. 175 of the children ate a hot lunch. How many children did not eat a hot lunch?

190 − 175 = 15

15 children

E Tree cutters were cutting down pine trees and maple trees. They loaded 185 pine logs and 190 maple logs onto railway cars. How many logs in all were loaded?

185 + 190 = 375

375 logs

F At the automobile races there were 1490 men. 2100 persons went to the races. How many were women?

2100 − 1490 = 610

610 women

Part 6 continues on the next page.

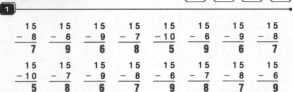

G Maria did 97 subtraction problems. Then she did some more problems. Now she has done 102 problems. How many more problems did she do?

$102 - 97 = 5$

5 problems

H Last summer Mike and Holly counted 158 crows and 190 sparrows. How many birds did they count?

$158 + 190 = 348$

348 birds

I A men's shop had 1500 shirts on sale. The shop sold 1389 of the shirts. How many shirts does the shop have now?

$1500 - 1389 = 111$

111 shirts

J Our family moved to Bellwood. We had 143 large boxes and 180 small boxes. How many boxes did we have?

$143 + 180 = 323$

323 boxes

7

A	B	C	D	E
5000	3400	7400	5100	6000
− 386	− 280	− 1648	− 243	− 5940
4614	3120	5752	4857	60

F	G	H	I	J
3100	5000	6300	1000	3010
− 264	− 4138	− 680	− 349	− 999
2836	862	5620	651	2011

Test + Facts + Problems + Bonus = TOTAL

1

15	15	15	15	15	15	15	15
− 7	− 6	− 8	− 10	− 9	− 6	− 10	− 8
8	9	7	5	6	9	5	7
15	15	15	15	15	15	15	15
− 9	− 7	− 6	− 9	− 8	− 7	− 9	− 6
6	8	9	6	7	8	6	9

2

A 14 {9, 5}: 14 − 9 = 5; 14 − 5 = 9

B 13 {9, 4}: 13 − 9 = 4; 13 − 4 = 9

C 17 {9, 8}: 17 − 9 = 8; 17 − 8 = 9

D 16 {9, 7}: 16 − 9 = 7; 16 − 7 = 9

3

17	14	14	16	17	12	13	16
− 8	− 5	− 7	− 8	− 8	− 9	− 4	− 9
9	9	7	8	9	3	9	7
16	14	17	12	14	16	16	14
− 7	− 9	− 8	− 6	− 7	− 7	− 9	− 5
9	5	9	6	7	9	7	9

4

11	13	16	11	13	13	11	14
− 6	− 8	− 7	− 8	− 4	− 5	− 7	− 5
5	5	9	3	9	8	4	9
11	13	11	9	11	9	14	13
− 3	− 6	− 5	− 2	− 4	− 4	− 8	− 7
8	7	6	7	7	5	6	6

Part 4 continues on the next page.

6	11	9	13	11	8	9	11
− 2	− 7	− 5	− 6	− 6	− 3	− 6	− 8
4	4	4	7	5	5	3	3
7	11	9	8	11	8	14	12
− 3	− 2	− 3	− 2	− 9	− 5	− 6	− 7
4	9	6	6	2	3	8	5

5

A Mrs. Oliver is 42 years old. Mr. Oliver is 60 years old. How many years younger is Mrs. Oliver?

$60 - 42 = 18$

18 years

B There were two hippopotamuses. The smaller one weighed 1342 kilograms. The larger hippopotamus weighed 1520 kilograms. How many kilograms lighter is the smaller hippopotamus?

$1520 - 1342 = 178$

178 kilograms

C The Ace Trucking Company has two trucks. The first truck was driven 1480 kilometers. The second truck was driven 1520 kilometers. How many kilometers in all were the trucks driven?

$1480 + 1520 = 3000$

3000 kilometers

D Two bridges are in nearby cities. The first bridge is 426 meters long. The second bridge is 510 meters long. How much longer is the second bridge than the first bridge?

$510 - 426 = 84$

84 meters

E Elisa picked 142 carrots from her garden. Her sister Melissa picked 180 carrots from her garden. How many carrots did both sisters pick from the gardens?

$142 + 180 = 322$

322 carrots

6

A Westchester had 438 buses. Then the city bought some more buses. Now there are 466 buses. How many more buses did the city buy?

$466 - 438 = 28$

28 buses

B There are 18 red wires and 24 black wires in the TV set. How many wires are there?

$18 + 24 = 42$

42 wires

C 1420 girls went to a parade. 1700 children went to the parade. How many boys went to the parade?

$1700 - 1420 = 280$

280 boys

D Mrs. Garcia ordered 164 bottles for her shop. 98 of them were broken. How many of the bottles were not broken?

$164 - 98 = 66$

66 bottles

E A factory built 2145 cars that had radios and 3190 cars that did not have radios. How many cars in all did the factory build?

$2145 + 3190 = 5335$

5335 cars

F There were 95 people in a restaurant. Later, more people came to the restaurant. Then there were 120 people in the restaurant. How many people came later?

$120 - 95 = 25$

25 people

Part 6 continues on the next page.

Lesson 58 (continued)

G 1500 runners were asked to try a new kind of running shoe. 864 runners did not like them. How many runners liked the new kind of shoe?

1500 − 864 = 636

636 runners

H In an exercise class 124 girls are jumping rope. 150 girls are swimming. How many girls are in the class?

124 + 150 = 274

274 girls

I 148 people were watching Eric shoot his bow and arrow. 52 more people joined them. How many people were watching Eric then?

148 + 52 = 200

200 people

J On Monday 1405 boxes of blueberries were for sale at a market. By Friday 1289 boxes of blueberries had been sold. How many boxes were left for the market to sell?

1405 − 1289 = 116

116 boxes

7

	A	B	C	D	E
	4600	1000	7043	4100	6000
−	2780	386	5239	3683	320
	1820	**614**	**1804**	**417**	**5680**

	F	G	H	I	J
	5310	4000	2400	7100	4100
−	4298	3369	908	6936	62
	1012	**631**	**1492**	**164**	**4038**

Lesson 59

Facts + Problems + Bonus = TOTAL

1

13 − 9 = **4**	13 − 6 = **7**	16 − 7 = **9**	12 − 6 = **6**	12 − 9 = **3**	14 − 7 = **7**	14 − 5 = **9**	14 − 9 = **5**
18 − 9 = **9**	15 − 6 = **9**	15 − 9 = **6**	13 − 4 = **9**	13 − 9 = **4**	17 − 8 = **9**	17 − 9 = **8**	15 − 6 = **9**

2

A 8: 2, [6] — 8 − 2 = 6; 8 − 6 = 2

B 6: 2, [4] — 6 − 2 = 4; 6 − 4 = 2

3

6 − 4 = **2**	11 − 7 = **4**	13 − 5 = **8**	11 − 8 = **3**	8 − 6 = **2**	13 − 4 = **9**	15 − 6 = **9**	11 − 6 = **5**
13 − 6 = **7**	11 − 4 = **7**	13 − 5 = **8**	8 − 6 = **2**	11 − 6 = **5**	11 − 3 = **8**	13 − 8 = **5**	6 − 4 = **2**
15 − 8 = **7**	11 − 2 = **9**	13 − 6 = **7**	11 − 5 = **6**	8 − 6 = **2**	11 − 2 = **9**	11 − 7 = **4**	13 − 5 = **8**
9 − 2 = **7**	13 − 6 = **7**	6 − 4 = **2**	15 − 7 = **8**	15 − 9 = **6**	11 − 8 = **3**	15 − 8 = **7**	15 − 6 = **9**

4

A Two camels were carrying food across the desert. A brown camel carried 126 kilograms of food. A gray camel carried 140 kilograms of food. How many kilograms lighter was the food that the brown camel carried?

140 − 126 = 14

14 kilograms

Part 4 continues on the next page.

Lesson 59 (continued)

B Steve and Kelly collect gold coins. Steve has 39 coins. Kelly has 46 coins. How many fewer coins does Steve have?

46 − 39 = 7

7 coins

C Bart and Roy raise chickens. Bart has 47 chickens and Roy has 92 chickens. How many chickens in all do they have?

47 + 92 = 139

139 chickens

D A card shop sold 1400 get-well cards and 1900 birthday cards last month. How many more birthday cards did the shop sell than get-well cards?

1900 − 1400 = 500

500 birthday cards

E Ian saw many boats on the lake as he rode his bicycle. He saw 85 sailboats and 29 motorboats. How many boats in all did Ian see?

85 + 29 = 114

114 boats

F Cathy has 147 joke books. Joe has 210 joke books. How many more joke books does Joe have than Cathy?

210 − 147 = 63

63 joke books

5

A The Happy Company has 90 pieces of pipe. 72 pieces of pipe are straight. How many pieces of pipe are not straight?

90 − 72 = 18

18 pieces

B We went walking in the woods. We saw 143 field mice and 180 rabbits. How many animals did we see?

143 + 180 = 323

323 animals

Part 5 continues on the next page.

Lesson 59 (continued)

C Mr. Banaka is a champion swimmer. He earned 1842 points for his swimming skill during the first hour of a contest. Then he got 88 more points for diving. How many points in all did Mr. Banaka get?

1842 + 88 = 1930

1930 points

D A soccer team has won 83 games in the last four years. The team has played 90 games. How many games has the team lost?

90 − 83 = 7

7 games

E A hotel bought 1444 light bulbs. 98 of them were broken. How many light bulbs were not broken?

1444 − 98 = 1346

1346 bulbs

F 136 people were riding horses. 96 people got off of their horses. How many people are still on their horses?

136 − 96 = 40

40 people

G A book shop sold 1826 books in December. The store sold 1900 books in January. How many books did the shop sell?

1826 + 1900 = 3726

3726 books

H 45 cows are in the barn. 90 cows are not in the barn. How many cows are there in all?

45 + 90 = 135

135 cows

Part 5 continues on the next page.

Lesson 59 (continued)

I When a cooking school opened, 148 people signed up to take classes. Other people came later to sign up for classes. Then there were 190 students in the cooking school. How many people came later?

[42] people

$190 - 148 = 42$ (borrow: 8, 1)

6

A. 3100 − 86 = 3014 (0 9 1)	B. 7000 − 230 = 6770 (6 9 1)	C. 4000 − 8 = 3992 (3 9 9 1)	D. 1000 − 20 = 980 (9 1)	E. 6200 − 4936 = 1264 (5 11 9 1)
F. 1000 − 923 = 77 (9 9 1)	G. 5100 − 486 = 4614 (4 10 9 1)	H. 2000 − 180 = 1820 (1 9 1)	I. 7200 − 943 = 6257 (6 11 9 1)	J. 6010 − 4798 = 1212 (5 9 10 1)

Lesson 60

Test + Facts + Problems + Bonus = TOTAL

1

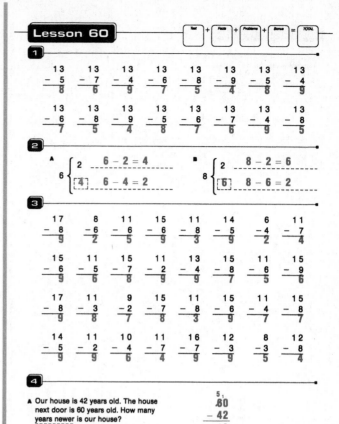

13 − 5 = 8	13 − 7 = 6	13 − 4 = 9	13 − 6 = 7	13 − 8 = 5	13 − 9 = 4	13 − 5 = 8	13 − 4 = 9
13 − 6 = 7	13 − 8 = 5	13 − 9 = 4	13 − 5 = 8	13 − 6 = 7	13 − 7 = 6	13 − 4 = 9	13 − 8 = 5

2

A 6 { 2, [4] — 6 − 2 = 4; 6 − 4 = 2

B 8 { 2, [6] — 8 − 2 = 6; 8 − 6 = 2

3

17 − 8 = 9	8 − 6 = 2	11 − 6 = 5	15 − 6 = 9	11 − 8 = 3	14 − 5 = 9	6 − 4 = 2	11 − 7 = 4
15 − 6 = 9	11 − 5 = 6	15 − 7 = 8	11 − 2 = 9	13 − 4 = 9	15 − 8 = 7	11 − 6 = 5	15 − 9 = 6
17 − 8 = 9	11 − 3 = 8	9 − 2 = 7	15 − 7 = 8	11 − 8 = 3	15 − 6 = 9	11 − 4 = 7	15 − 8 = 7
14 − 5 = 9	11 − 2 = 9	10 − 4 = 6	11 − 7 = 4	16 − 7 = 9	12 − 3 = 9	8 − 3 = 5	12 − 8 = 4

4

A Our house is 42 years old. The house next door is 60 years old. How many years newer is our house?

[18] years

$60 - 42 = 18$ (borrow: 5, 1)

Part 4 continues on the next page.

Lesson 60 (continued)

B There are 46 white turkeys and 90 gray turkeys on a farm. How many more gray turkeys are there than white ones?

[44] gray turkeys

$90 - 46 = 44$ (borrow: 8, 1)

C A carnival sold 94 balloons on the first day. On the second day the carnival sold 120 balloons. How many balloons did the carnival sell in all?

[214] balloons

$94 + 120 = 214$

D There were two huge statues in a park. One weighed 1340 kilograms. The other weighed 1700 kilograms. How many kilograms did the statues weigh in all?

[3040] kilograms

$1340 + 1700 = 3040$

E A fruit grower sent 1400 boxes of lemons and 1900 boxes of oranges to a market. How many more boxes of oranges were sent than lemons?

[500] boxes

$1900 - 1400 = 500$

5

A Mr. Jenssen bought an old chair that was 37 years old. Mr. Merino bought a sofa that was 50 years old. How many years older was the sofa?

[13] years

$50 - 37 = 13$ (borrow: 4, 1)

B In a parade we saw 150 police officers riding horses and 150 cowboys riding horses. How many people in all were riding horses?

[300] people

$150 + 150 = 300$

Part 5 continues on the next page.

Lesson 60 (continued)

C Our city has 1960 green taxis and 3045 red taxis. How many taxis does our city have?

[5005] taxis

$1960 + 3045 = 5005$

D Louis and Girard were writing a book. In May they wrote 175 pages. In June they wrote 190 pages. How many pages did they write in May and June?

[365] pages

$175 + 190 = 365$

E A huge oak desk weighs 124 kilograms. A brass lamp on the desk weighs 48 kilograms. How much do the desk and lamp weigh altogether?

[172] kilograms

$124 + 48 = 172$

F Our basketball team won 47 games. The team played 52 games in all. How many games did the team lose?

[5] games

$52 - 47 = 5$ (borrow: 4, 1)

G 42 trains went through the tunnel on Sunday. 38 trains went through the tunnel on Monday. How many trains went through the tunnel?

[80] trains

$42 + 38 = 80$

H 154 passengers were on a plane that was going to China. In Los Angeles 99 people got off. How many people did not get off?

[55] people

$154 - 99 = 55$ (borrow: 0 4, 1)

I A men's shop sold 85 plain neckties and 139 striped neckties. How many neckties did the shop sell?

[224] neckties

$85 + 139 = 224$

Part 5 continues on the next page.

Lesson 60 (continued)

J Mr. Johnson has a newsstand. He sold 138 newspapers in the morning. Then he sold some more newspapers in the afternoon. Now he has sold 150 newspapers. How many newspapers did he sell in the afternoon?

150 − 138 = 12

12 newspapers

6

A 4000 − 1236 = 2764	B 8100 − 26 = 8074	C 4030 − 594 = 3436	D 6000 − 4123 = 1877	E 5200 − 186 = 5014
F 3100 − 84 = 3016	G 1000 − 90 = 910	H 4000 − 1555 = 2445	I 3600 − 280 = 3320	J 4126 − 3996 = 130

Lesson 61

Facts + Problems + Bonus = TOTAL

1

13 − 6 = 7	13 − 4 = 9	13 − 9 = 4	13 − 5 = 8	13 − 6 = 7	13 − 8 = 5	13 − 7 = 6	13 − 9 = 4
13 − 6 = 7	13 − 4 = 9	13 − 7 = 6	13 − 5 = 8	13 − 8 = 5	13 − 9 = 4	13 − 6 = 7	13 − 4 = 9

2

6 − 4 = 2	11 − 7 = 4	17 − 8 = 9	11 − 6 = 5	15 − 7 = 8	14 − 5 = 9	8 − 6 = 2	11 − 8 = 3
11 − 4 = 7	16 − 7 = 9	11 − 9 = 2	15 − 8 = 7	14 − 6 = 8	13 − 4 = 9	8 − 6 = 2	11 − 5 = 6
11 − 3 = 8	14 − 5 = 9	6 − 4 = 2	11 − 8 = 3	15 − 6 = 9	11 − 7 = 4	15 − 7 = 8	11 − 4 = 7
17 − 8 = 9	15 − 8 = 7	9 − 4 = 5	8 − 3 = 5	13 − 4 = 9	8 − 5 = 3	12 − 5 = 7	15 − 6 = 9

3

A There are 170 birds in a zoo. There are 200 fish in the zoo. How many fewer birds than fish are in the zoo?

200 − 170 = 30

30 birds

B We saw lots of squirrels and rabbits. We counted 143 squirrels and 180 rabbits. How many animals did we see?

143 + 180 = 323

323 animals

Part 3 continues on the next page.

Lesson 61 (continued)

C The cook let us choose cake or ice cream for dessert. 145 kids ate cake and 200 kids ate ice cream. How many more kids ate ice cream than cake?

200 − 145 = 55

55 kids

D Iris taught 36 children how to jump rope. Christina taught 40 children how to jump rope. How many children learned how to jump rope?

36 + 40 = 76

76 children

E Chris has 47 rocks. Judd has 82 rocks. How many more rocks does Judd have than Chris?

82 − 47 = 35

35 rocks

F There are 90 bottles of soda pop. 72 of the bottles are full. How many bottles are not full?

90 − 72 = 18

18 bottles

G 1350 railway cars are full. 1500 railway cars are empty. How many railway cars are there in all?

1350 + 1500 = 2850

2850 railway cars

H The Fireside Bookshop sold 1826 books in December. The shop sold 1900 books in January. How many books did the shop sell?

1826 + 1900 = 3726

3726 books

I A hockey team lost 42 games. It played 70 games in all. How many games did the hockey team win?

70 − 42 = 28

28 games

Part 3 continues on the next page.

Lesson 61 (continued)

J Fran ran 37 kilometers this week. Carol ran 40 kilometers this week. How many kilometers did Fran and Carol run this week?

37 + 40 = 77

77 kilometers

K Louise made 49 small cakes with strawberry frosting. Then she made some cakes with chocolate frosting. She baked 72 cakes in all. How many had chocolate frosting?

72 − 49 = 23

23 cakes

L Ann read 167 pages of a book. On her day off she read some more. Now Ann has read 210 pages. How many pages did Ann read on her day off?

210 − 167 = 43

43 pages

4

A 3010 − 826 = 2184	B 5200 + 380 = 5580	C 6000 − 84 = 5916	D 4106 − 3289 = 817	E 5006 − 95 = 4911
F 8105 − 7999 = 106	G 5010 − 4364 = 646	H 8246 − 7580 = 666	I 9000 − 4280 = 4720	J 6400 + 287 = 6687
K 3200 − 1800 = 1400	L 5130 − 4830 = 300	M 9042 + 186 = 9228	N 7036 − 189 = 6847	O 1000 − 285 = 715

Lesson 62

Facts + Problems + Bonus = TOTAL

1

A 17: 9 — 17 − 9 = 8; [8] — 17 − 8 = 9

B 14: 9 — 14 − 9 = 5; [5] — 14 − 5 = 9

C 15: 9 — 15 − 9 = 6; [6] — 15 − 6 = 9

D 16: 9 — 16 − 9 = 7; [7] — 16 − 7 = 9

2

15	17	12	10	16	14	14	15
− 9	− 8	− 6	− 5	− 7	− 7	− 9	− 6
6	9	6	5	9	7	5	9
15	14	14	14	17	17	16	14
− 7	− 5	− 6	− 7	− 8	− 9	− 7	− 5
8	9	8	7	9	8	9	9

3

14	17	9	13	8	15	6	11
− 6	− 8	− 4	− 8	− 6	− 7	− 4	− 4
8	9	5	5	2	8	2	7
15	13	15	8	11	6	14	11
− 7	− 7	− 6	− 6	− 8	− 4	− 5	− 3
8	6	9	2	3	2	9	8
15	9	6	13	15	12	13	11
− 8	− 5	− 4	− 6	− 8	− 3	− 7	− 2
7	4	2	7	7	9	6	9
13	14	13	8	13	17	15	13
− 4	− 9	− 5	− 5	− 8	− 8	− 7	− 6
9	5	8	3	5	9	8	7

Lesson 62 (continued)

4

A 64 people walked around the ladder. 140 people walked under the ladder. How many more people walked under the ladder than around the ladder?

[76] people — 140 − 64 = 76

B In one week 148 new cars and 473 old cars were washed at the Happy Car Wash. How many cars in all were washed?

[621] cars — 148 + 473 = 621

C At a camp 3802 oranges were eaten in one week. 1648 apples were also eaten. How many more oranges than apples were eaten?

[2154] oranges — 3802 − 1648 = 2154

D Andy put 2509 boxes on trucks. Randy put 3108 boxes on trucks. How many fewer boxes did Andy put on trucks than Randy?

[599] boxes — 3108 − 2509 = 599

E Workers chopped down 185 oak trees and 190 pine trees. How many trees were chopped down?

[375] trees — 185 + 190 = 375

F Last spring a shop had a sale on kites. The shop sold 158 red kites and 190 yellow kites. How many kites did the shop sell?

[348] kites — 158 + 190 = 348

Part 4 continues on the next page.

Lesson 62 (continued)

G Lauren planted 97 flowers. Then she planted some more flowers. Now 102 flowers are planted. How many more flowers did she plant?

[5] flowers — 102 − 97 = 5

H A newspaper stand received 1444 magazines. 98 of the magazines had torn pages. How many magazines did not have torn pages?

[1346] magazines — 1444 − 98 = 1346

I 148 people were ice-skating. Later, more people came to skate. Now there are 190 skaters. How many people came late to skate?

[42] people — 190 − 148 = 42

J Phil danced for 148 minutes. Then he danced for another 52 minutes. How many minutes did he dance in all?

[200] minutes — 148 + 52 = 200

K Workers cleaned 18 airplanes. They cleaned 24 boats. How many things did the workers clean?

[42] things — 18 + 24 = 42

L An office had 1405 packages of paper. The office used up 1289 packages. How many packages of paper were left?

[116] packages — 1405 − 1289 = 116

Lesson 62 (continued)

5

A 5000 − 4836 = 164	**B** 6200 + 886 = 7086	**C** 5100 − 284 = 4816	**D** 3200 − 1964 = 1236
E 1000 − 20 = 980	**F** 3400 − 1950 = 1450	**G** 5010 − 4286 = 724	**H** 3825 + 182 = 4007
I 9306 − 829 = 8477	**J** 3000 − 1280 = 1720		

Lesson 63

Test + Facts + Problems + Bonus = TOTAL

1

14 − 5 = 9	14 − 9 = 5	14 − 7 = 7	17 − 8 = 9	17 − 9 = 8	16 − 8 = 8	15 − 6 = 9	13 − 4 = 9
18 − 9 = 9	14 − 7 = 7	16 − 7 = 9	14 − 5 = 9	12 − 6 = 6	17 − 8 = 9	17 − 9 = 8	16 − 7 = 9

2

14 − 5 = 9	11 − 7 = 4	13 − 6 = 7	11 − 9 = 2	16 − 7 = 9	15 − 7 = 8	11 − 8 = 3	11 − 6 = 5
13 − 7 = 6	13 − 4 = 9	15 − 8 = 7	11 − 4 = 7	13 − 5 = 8	12 − 3 = 9	13 − 4 = 9	11 − 3 = 8
13 − 8 = 5	17 − 8 = 9	11 − 5 = 6	13 − 7 = 6	15 − 6 = 9	11 − 7 = 4	13 − 5 = 8	11 − 4 = 7
8 − 3 = 5	11 − 2 = 9	8 − 6 = 2	11 − 8 = 3	6 − 4 = 2	14 − 5 = 9	8 − 3 = 5	9 − 4 = 5

3

A Martin and Frank were selling balloons. Martin sold 93 balloons. Frank sold 130 balloons. How many fewer balloons did Martin sell than Frank?
37 balloons
130 − 93 = 37

Part 3 continues on the next page.

Lesson 63 (continued)

B 425 people voted for Sam. 128 people did not vote for Sam. How many people voted in all?
553 people
425 + 128 = 553

C 195 people came to the meeting early. 24 people came late. How many people in all came to the meeting?
219 people
195 + 24 = 219

D Rob has 36 rabbits. Tony has 52 rabbits. How many more rabbits does Tony have than Rob?
16 rabbits
52 − 36 = 16

E There were 154 fish. 89 of the fish were swordfish. How many of the fish were not swordfish?
65 fish
154 − 89 = 65

F In a building 85 windows were closed. 17 windows were open. How many windows did the building have?
102 windows
85 + 17 = 102

G A sign outside an office building was 48 meters tall. It stood next to a pole that was 60 meters tall. How many meters taller was the pole?
12 meters
60 − 48 = 12

H In our city there are 548 brick buildings and 379 wooden buildings. How many buildings are in our city?
927 buildings
548 + 379 = 927

Part 3 continues on the next page.

Lesson 63 (continued)

I 143 planes landed late. 583 planes did not land late. How many planes landed in all?
726 planes
143 + 583 = 726

J 126 runners entered a race. Some more runners entered the race. Now there are 142 runners. How many more runners entered the race?
16 runners
142 − 126 = 16

K There were 1490 children at a picnic. There were 2100 people at the picnic. How many adults were at the picnic?
610 adults
2100 − 1490 = 610

L 1500 people went to a flower show. 864 people did not buy flowers and plants. How many people did buy flowers and plants?
636 people
1500 − 864 = 636

4

A 3200 − 864 = 2336	**B** 6400 − 4980 = 1420	**C** 6000 − 37 = 5963	**D** 1280 + 880 = 2160	**E** 7100 − 3489 = 3611
F 6010 − 999 = 5011	**G** 1000 − 640 = 360	**H** 1000 − 39 = 961	**I** 6100 − 436 = 5664	**J** 9500 − 3820 = 5680

Lesson 64

Test + Facts + Problems + Bonus = TOTAL

1

15 − 6 = 9	15 − 9 = 6	14 − 7 = 7	14 − 5 = 9	14 − 9 = 5	16 − 8 = 8	17 − 8 = 9	16 − 8 = 8
16 − 7 = 9	16 − 9 = 7	16 − 8 = 8	13 − 4 = 9	13 − 9 = 4	15 − 6 = 9	15 − 9 = 6	16 − 8 = 8

2

11 − 5 = 6	14 − 5 = 9	6 − 4 = 2	15 − 7 = 8	11 − 8 = 3	13 − 5 = 8	17 − 8 = 9	11 − 7 = 4
13 − 8 = 5	15 − 7 = 8	13 − 6 = 7	11 − 4 = 7	15 − 8 = 7	11 − 6 = 5	8 − 6 = 2	11 − 3 = 8
13 − 7 = 6	10 − 3 = 7	11 − 2 = 9	13 − 5 = 8	16 − 7 = 9	15 − 8 = 7	6 − 4 = 2	13 − 8 = 5
14 − 9 = 5	13 − 6 = 7	9 − 4 = 5	8 − 6 = 2	13 − 7 = 6	12 − 4 = 8	13 − 5 = 8	12 − 8 = 4

3

A Art is 48 years old. Troy is 70 years old. How many years younger is Art?
22 years
70 − 48 = 22

B A spaceship went around the earth 34 times the first week. During the second week the spaceship went around the earth 30 times. How many times did it go around the earth?
64 times
34 + 30 = 64

Part 3 continues on the next page.

Lesson 64 (continued)

C A cook used 185 brown eggs and 480 white eggs. How many eggs did the cook use?

665 eggs

185
+ 480
665

D A city has 148 old buses and 172 new buses. How many more new buses than old buses are there?

24 buses

6,
172
− 148
24

E Carol is 146 centimeters tall. Lupe is 164 centimeters tall. How many centimeters taller is Lupe?

18 centimeters

5,
164
− 146
18

F A new shop gave away 86 red pens and 90 blue pens. How many pens did the shop give away?

176 pens

86
+ 90
176

G Nancy has a dog that used to weigh 108 kilograms. The dog has gained 24 kilograms. How much does it weigh now?

132 kilograms

108
+ 24
132

H A small car weighs 1248 kilograms. A racing car weighs 1600 kilograms. How much heavier is the racing car?

352 kilograms

5 9,
1600
− 1248
352

I We had 164 nuts. 98 of the nuts had shells. How many of the nuts did not have shells?

66 nuts

0'5,
164
− 98
66

Part 3 continues on the next page.

Lesson 64 (continued)

J Martha Rainwater counted 1463 ducks on the lake and 3652 ducks in the air. How many ducks did she count?

5115 ducks

1463
+ 3652
5115

K 184 girls were playing checkers. 176 boys were playing checkers. How many children were playing checkers?

360 children

184
+ 176
360

L At the beginning of the week Kurt had driven his car 945 kilometers. At the end of the week Kurt had driven 1200 kilometers. How many kilometers did Kurt drive his car during the week?

255 kilometers

0'1 9,
1200
− 945
255

Lesson 65

Test + Problems + Bonus = TOTAL

1

6 − 4 2	13 − 9 4	17 − 8 9	8 − 6 2	10 − 4 6	13 − 5 8	11 − 2 9	15 − 7 8
11 − 8 3	16 − 9 7	11 − 6 5	13 − 7 6	10 − 6 4	11 − 4 7	12 − 9 3	13 − 4 9
8 − 2 6	14 − 5 9	16 − 7 9	10 − 2 8	13 − 6 7	11 − 9 2	9 − 5 4	7 − 3 4
13 − 8 5	16 − 9 7	11 − 3 8	14 − 9 5	8 − 5 3	15 − 6 9	7 − 4 3	12 − 4 8
9 − 3 6	15 − 8 7	10 − 3 7	9 − 4 5	9 − 7 2	17 − 9 8	10 − 7 3	12 − 3 9
14 − 6 8	8 − 3 5	15 − 9 6	9 − 2 7	11 − 5 6	10 − 2 8	14 − 8 6	11 − 7 4
12 − 4 8	5 − 3 2	11 − 7 4	12 − 8 4	12 − 5 7	9 − 6 3	12 − 6 6	10 − 4 6

2

A 148 motorboats were on the river. 162 sailboats were on the river. How many more sailboats than motorboats were on the river?

14 sailboats

5,
162
− 148
14

Part 2 continues on the next page.

Lesson 65 (continued)

B A hospital had 2365 clean sheets. It had 90 dirty sheets. How many sheets altogether did the hospital have?

2455 sheets

2365
+ 90
2455

C Leslie has worked 800 hours so far this year. Holly has worked 1250 hours. How many more hours has Holly worked?

450 hours

1250
− 800
450

D A jar had 438 coins in it. Some more coins were put in the jar. Now there are 466 coins in the jar. How many more coins were put in the jar?

28 coins

5,
466
− 438
28

E 5200 men watched the play. 4800 women watched the play. How many more men watched the play than women?

400 men

4,
5200
− 4800
400

F Before eating their picnic lunch, 86 people bought milk. 14 people bought milk after lunch. How many people bought milk?

100 people

86
+ 14
100

G In our high school 124 girls are learning how to play tennis. 150 other girls are learning how to play volleyball. How many girls are learning how to play those games?

274 girls

124
+ 150
274

Part 2 continues on the next page.

Lesson 65 (continued)

H Gloria is 137 centimeters tall. Ginger is 143 centimeters tall. How many centimeters taller is Ginger?

6 centimeters

143
− 137
6

I A garden had 860 roses. 290 of the roses were red. How many roses were not red?

570 roses

860
− 290
570

J Last week Mr. Kelly fixed 24 cars. This week he fixed 30 cars. How many cars has he fixed altogether?

54 cars

24
+ 30
54

K Mrs. Owens is a taxi driver. Last week she drove 436 kilometers. This week she drove 544 kilometers. How many more kilometers did she drive this week?

108 kilometers

544
− 436
108

L A pet shop sold 46 puppies and 70 kittens last month. How many animals did the pet shop sell?

116 animals

46
+ 70
116

M There were 199 students in our school. Some more students came to our school. Now there are 215 students in our school. How many students came to our school?

16 students

215
− 199
16

N A machine has 1584 parts. 990 parts are not new. How many parts are new?

594 parts

1584
− 990
594

Lesson 65 (continued)

3

A	B	C	D
5000	5300	1000	5100
− 4820	− 1849	− 87	+ 826
180	4451	913	5926

E	F	G	H
9400	7000	4500	6100
− 826	− 23	+ 890	− 87
8574	6977	5390	6013

I	J
3040	5200
− 2986	− 980
54	4220

Transition Lesson 8

BONUS

1

A	B	C	D	E	F
264	538	492	751	892	376

2

A	B	C	D	E	F
406	308	728	507	346	205

3

A 7 { 4: 7 − 4 = 3; 3: 7 − 3 = 4

B 9 { 2: 9 − 2 = 7; 7: 9 − 7 = 2

C 3 { 2: 3 − 2 = 1; 1: 3 − 1 = 2

D 5 { 3: 5 − 3 = 2; 2: 5 − 2 = 3

Transition Lesson 8 (continued)

4

A 4 { 2: 2 + 2 = 4; 2: 4 − 2 = 2

B 10 { 5: 5 + 5 = 10; 5: 10 − 5 = 5

C 18 { 9: 9 + 9 = 18; 9: 18 − 9 = 9

D 6 { 3: 3 + 3 = 6; 3: 6 − 3 = 3

5

10	6	6	18	6	8	7	18
− 5	− 3	− 1	− 9	− 3	− 1	− 1	− 9
5	3	5	9	3	7	6	9

2	10	6	8	18	10	10	7
− 1	− 5	− 3	− 1	− 9	− 5	− 1	− 1
1	5	3	7	9	5	9	6

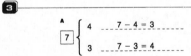

Transition Lesson 25

BONUS

1

A	B	C	D	E
4285	3624	8153	9284	7295

2

A	B	C	D	E	F
4056	5207	8024	2902	4908	7067

3

	A	B	C	D	E
	435	629	5734	7765	867
	− 414	− 511	− 5124	− 7160	− 450
	21	118	610	605	417

4

A
7 { 4 — 7 − 4 = 3
7 { 3 — 7 − 3 = 4

B
9 { 2 — 9 − 2 = 7
9 { 7 — 9 − 7 = 2

C
3 { 2 — 3 − 2 = 1
3 { 1 — 3 − 1 = 2

D
5 { 3 — 5 − 3 = 2
5 { 2 — 5 − 2 = 3

Transition Lesson 25 (continued)

5

A 60 − 1 = 59   B 70 − 1 = 69   C 20 − 1 = 19

D 40 − 1 = 39   E 30 − 1 = 29   F 50 − 1 = 49

G 90 − 1 = 89   H 80 − 1 = 79

6

	A	B	C	D
	2, 1,	7, 7,	2, 8,	2, 3,
	3420	8486	3092	3240
	− 1801	− 3768	− 1146	− 1739
	1619	**4718**	**1946**	**1501**

7

	A	B	C	D	E
	4,	2,	7,		3,
	752	3493	3480	70	2400
	− 24	− 610	− 51	+ 38	− 90
	728	**2883**	**3429**	**108**	**2310**

	F	G	H	I	J
			1,	6,	
	3528	738	9248	7251	3846
	+ 528	+ 8	− 1068	− 601	+ 3218
	4056	**746**	**8180**	**6650**	**7064**

	K	L	M	N	O
	5,			7,	5,
	460	5284	9328	4822	8605
	− 355	+ 6	− 16	− 142	− 7150
	105	**5290**	**9312**	**4680**	**1455**

Mastery Test Review—Lesson 12

1

2	5	7	4	8	6	8	4
− 0	− 0	− 1	− 0	− 1	− 0	− 0	− 1
2	**5**	**6**	**4**	**7**	**6**	**8**	**3**
3	8	6	4	10	3	6	2
− 0	− 1	− 0	− 0	− 1	− 1	− 0	− 0
3	**7**	**6**	**4**	**9**	**2**	**6**	**2**

Mastery Test Review—Lesson 15

1

12	11	14	15	16	13	14	19
− 6	− 10	− 7	− 10	− 8	− 10	− 7	− 10
6	**1**	**7**	**5**	**8**	**3**	**7**	**9**
18	14	18	12	8	8	16	12
− 9	− 10	− 10	− 6	− 4	− 0	− 8	− 10
9	**4**	**8**	**6**	**4**	**8**	**8**	**2**
12	14	16	16	18	6	6	10
− 6	− 7	− 10	− 8	− 9	− 3	− 1	− 1
6	**7**	**6**	**8**	**9**	**3**	**5**	**9**

2

16	16	14	14	14	14	18	12
− 10	− 8	− 7	− 10	− 8	− 6	− 9	− 6
6	**8**	**7**	**4**	**6**	**8**	**9**	**6**
12	16	14	14	12	15	14	10
− 10	− 8	− 6	− 8	− 6	− 10	− 8	− 5
2	**8**	**8**	**6**	**6**	**5**	**6**	**5**
14	14	14	14	8	6	18	16
− 6	− 10	− 7	− 8	− 4	− 3	− 9	− 8
8	**4**	**7**	**6**	**4**	**3**	**9**	**8**

Mastery Test Review—Lesson 22

1

	A	B	C	D	E
	8,	5,	7,	8,	6,
	396	607	848	942	7875
	− 248	− 497	− 670	− 180	− 2965
	148	**110**	**178**	**762**	**4910**

	F	G	H	I	J
	2,	2,	8,	7,	2,
	932	3643	5990	825	3046
	− 826	− 1840	− 4189	− 160	− 1506
	106	**1803**	**1801**	**665**	**1540**

2

	A	B	C	D	E
	8,	8,	5,	6,	6,
	492	903	681	7043	76
	− 286	− 791	− 491	− 5110	− 38
	206	**112**	**190**	**1933**	**38**

	F	G	H	I	J
	4,	5,	5,	7,	6,
	3500	9601	6824	584	745
	− 1350	− 490	− 1904	− 466	− 80
	2150	**9111**	**4920**	**118**	**665**

Mastery Test Review—Lesson 24

1

A
12 { 4 — 12 − 4 = 8
12 { 8 — 12 − 8 = 4

B
12 { 2 — 12 − 2 = 10
12 { 10 — 12 − 10 = 2

C
12 { 3 — 12 − 3 = 9
12 { 9 — 12 − 9 = 3

D
12 { 5 — 12 − 5 = 7
12 { 7 — 12 − 7 = 5

Mastery Test Review Lesson 24 (continued)

2

12	12	12	12	12	12	12	12
− 5	− 8	− 7	− 9	− 6	− 7	− 8	− 9
7	**4**	**5**	**3**	**6**	**5**	**4**	**3**

12	12	12	12	12	12	12	12
− 4	− 9	− 5	− 8	− 3	− 7	− 6	− 10
8	**3**	**7**	**4**	**9**	**5**	**6**	**2**

Mastery Test Review—Lesson 30

1

A 12 { 5, **7** } **12 − 5 = 7**

B 7 { 6, **1** } **7 − 6 = 1**

C **10** { 9, 1 } **9 + 1 = 10**

D **10** { 7, 3 } **7 + 3 = 10**

E 7 { 2, **5** } **7 − 2 = 5**

F **11** { 10, 1 } **10 + 1 = 11**

2

A 9 { 8, **1** } **9 − 8 = 1**

B **14** { 8, 6 } **8 + 6 = 14**

Part 2 continues on the next page.

Mastery Test Review Lesson 30 (continued)

C **12** { 3, 9 } **3 + 9 = 12**

D 3 { 1, **2** } **3 − 1 = 2**

E **12** { 8, 4 } **8 + 4 = 12**

F 7 { 5, **2** } **7 − 5 = 2**

Mastery Test Review—Lesson 32

1

A	B	C	D	E	F
4006	5027	7004	2902	4098	7007

2

A	B	C	D	E	F
5009	9035	8024	7406	9004	8201

Mastery Test Review—Lesson 33

1

A	B	C	D	E
39,	49,	69,	49,	89,
4046	5024	3704	5029	904
−2185	−1853	−2198	−3641	−186
1861	**3171**	**1506**	**1388**	**718**

2

A	B	C	D	E
49,	2,	49,	6,	29,
502	308	5049	7082	3068
−185	−157	− 768	−3140	− 184
317	**151**	**4281**	**3942**	**2884**

Mastery Test Review—Lesson 37

1

16	11	14	12	16	13	17	15
− 9	− 9	− 9	− 9	− 9	− 9	− 9	− 9
7	**2**	**5**	**3**	**7**	**4**	**8**	**6**

2

10	10	10	10	10	10	10	10
− 7	− 6	− 9	− 8	− 6	− 9	− 7	− 8
3	**4**	**1**	**2**	**4**	**1**	**3**	**2**

10	10	10	10	10	10	10	10
− 3	− 7	− 2	− 4	− 8	− 6	− 9	− 1
7	**3**	**8**	**6**	**2**	**4**	**1**	**9**

Mastery Test Review—Lesson 43

1

A Our school had a dog show for German shepherds and hunting dogs. There were 15 German shepherds in the show. 19 dogs had been brought to the show. How many hunting dogs were there?

4 **hunting dogs** — 19 − 15 = **4**

B In the morning 16 jet planes left New York. In the afternoon 74 small planes left New York. How many planes in all left New York?

90 **planes** — 16 + 74 = **90**

C There are 7 rosebushes in our garden. We have 4 red rosebushes. The rest are yellow. How many yellow rosebushes do we have?

3 **yellow rosebushes** — 7 − 4 = **3**

Part 1 continues on the next page.

Mastery Test Review Lesson 43 (continued)

D There are 14 clean cups on the shelf. There are 19 dirty cups in the sink. How many cups are there in all?

33 **cups** — 14 + 19 = **33**

E 14 women were in the play. 29 people were in the play. How many men were in the play?

15 **men** — 29 − 14 = **15**

F My aunt has 28 white flowers. She has 39 flowers in all. The rest of the flowers are pink. How many pink flowers does my aunt have?

11 **pink flowers** — 39 − 28 = **11**

2

A Bill painted 43 houses. He painted 52 stores. How many buildings did Bill paint?

95 **buildings** — 43 + 52 = **95**

B In the waiting room of a doctor's office there is a huge fish tank. There are 14 goldfish in the tank and the rest are sunfish. There are 30 fish in all in the tank. How many sunfish are in the tank?

16 **sunfish** — (2,) 30 − 14 = **16**

C There are 40 girls in the kindergarten of a school. There are 75 children in kindergarten. How many are boys?

35 **boys** — 75 − 40 = **35**

D Miss Manos is selling tickets. She sold 31 circus tickets this morning. This afternoon she sold 50 movie tickets. How many tickets in all did she sell?

81 **tickets** — 31 + 50 = **81**

Part 2 continues on the next page.

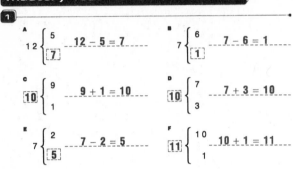

Mastery Test Review Lesson 43 (continued)

E Marcia Kabatie has entered many skating contests. 43 were ice-skating contests and the rest were roller-skating ones. She has taken part in 60 contests in all. How many roller-skating contests has she entered?

5 1
60
− 43
17

17 roller-skating contests

Mastery Test Review—Lesson 44

1

9	9	9	9	9	9	9	9
− 7	− 4	− 3	− 6	− 2	− 8	− 7	− 1
2	5	6	3	7	1	2	8
9	9	9	9	9	9	9	9
− 5	− 6	− 4	− 7	− 3	− 6	− 2	− 7
4	3	5	2	6	3	7	2

2

13	13	13	13	13	13	13	13
− 7	− 9	− 4	− 8	− 5	− 6	− 4	− 7
6	4	9	5	8	7	9	6
13	13	13	13	13	13	13	13
− 6	− 8	− 4	− 5	− 9	− 7	− 6	− 8
7	5	9	8	4	6	7	5

Mastery Test Review—Lesson 48

1

A There are 143 cabins up at the lake. This year more cabins were built. Now there are 160. How many more cabins were built at the lake?

5 1
160
− 143
17

17 cabins

B My dad's store was having a sale. There were 114 people in the store. Some more people came in. Then there were 142 people in the store. How many more people came into the store?

3 1
142
− 114
28

28 people

C Benji had a stamp collection. When the year began, he had 1436 stamps. During the year he got 148 stamps. How many stamps did Benji have at the end of the year?

1436
+ 148
1584

1584 stamps

D A meal was served at a banquet. There were 143 people sitting at the tables. Some more people arrived. Now there are 160 people at the tables. How many more people arrived?

5 1
160
− 143
17

17 people

E When the game started, there were 842 people watching it. During the game another 58 people began to watch it. How many people were watching the game altogether?

842
+ 58
900

900 people

Mastery Test Review Lesson 48 (continued)

2

A When the movie began, there were 146 people in the theater. By the end of the movie there were 206 people in the theater. How many people came into the theater after the movie started?

1 1
206
− 146
60

60 people

B Last summer we took photographs of a building that was being built. At that time the building had 58 floors. In winter the building was finished. It had 70 floors. How many floors were built after we took photographs?

6 1
70
− 58
12

12 floors

C 315 children took the bus to summer camp. Another 46 children took the train to camp. How many children went to summer camp in all?

315
+ 46
361

361 children

D At a beach there were 54 swimmers. 40 more swimmers came. How many swimmers were at the beach then?

54
+ 40
94

94 swimmers

E A dress shop had 23 dresses. Some more dresses came in a truck. Now there are 70 dresses in the shop. How many dresses came in the truck?

6 1
70
− 23
47

47 dresses

Mastery Test Review—Lesson 50

1

7	8	6	10	8	5	11	9
− 2	− 2	− 2	− 2	− 2	− 2	− 2	− 2
5	6	4	8	6	3	9	7
5	10	6	11	7	8	4	9
− 2	− 2	− 2	− 2	− 2	− 2	− 2	− 2
3	8	4	9	5	6	2	7

2

15	15	15	15	15	15	15	15
− 6	− 10	− 7	− 9	− 8	− 6	− 9	− 7
9	5	8	6	7	9	6	8
15	15	15	15	15	15	15	15
− 10	− 7	− 6	− 9	− 8	− 7	− 6	− 9
5	8	9	6	7	8	9	6

Mastery Test Review—Lesson 53

1

	A	B	C	D	E
	4 9 9 1	4 9 1	4 1	7 9 9 1	7 9 1
	5000	5000	5000	8000	8000
	− 3	− 30	− 300	− 4	− 40
	4997	4970	4700	7996	7960

	F	G	H	I	J
	2 9 9 1	2 1	9 9 1	9 1	
	3000	3000	1000	1000	1000
	− 8	− 800	− 7	− 20	− 400
	2992	2200	993	980	600

Mastery Test Review Lesson 53 (continued)

2

A. $\begin{array}{r} {}^{2\,9\,9\,1}\\ \not{3}000 \\ -\quad 8 \\ \hline 2992 \end{array}$
B. $\begin{array}{r} {}^{2\,1}\\ \not{3}000 \\ -\ 800 \\ \hline 2200 \end{array}$
C. $\begin{array}{r} {}^{2\,9\,1}\\ \not{3}000 \\ -\quad 80 \\ \hline 2920 \end{array}$
D. $\begin{array}{r} {}^{9\,1}\\ \not{1}000 \\ -\quad 20 \\ \hline 980 \end{array}$
E. $\begin{array}{r} {}^{9\,9\,1}\\ \not{1}000 \\ -\quad 2 \\ \hline 998 \end{array}$

F. $\begin{array}{r} {}^{3\,9\,9\,1}\\ \not{4}000 \\ -\ 245 \\ \hline 3755 \end{array}$
G. $\begin{array}{r} {}^{3\,9\,1}\\ \not{4}000 \\ -2450 \\ \hline 1550 \end{array}$
H. $\begin{array}{r} {}^{7\,9\,9\,1}\\ \not{8}000 \\ -\ 342 \\ \hline 7658 \end{array}$
I. $\begin{array}{r} {}^{7\,1}\\ \not{8}000 \\ -\ 300 \\ \hline 7700 \end{array}$
J. $\begin{array}{r} {}^{5\,9\,9\,1}\\ \not{6}000 \\ -4128 \\ \hline 1872 \end{array}$

Mastery Test Review—Lesson 54

1

A. 90 children went to a play. 46 of the children enjoyed the play. The rest did not. How many of the children did not enjoy the play?
$\begin{array}{r} {}^{8\,1}\\ \not{9}0 \\ -46 \\ \hline 44 \end{array}$

B. We bought some eggs. When we got home, 36 eggs were broken. 24 eggs were not broken. How many eggs in all did we buy?
$\begin{array}{r} 36 \\ +24 \\ \hline 60 \end{array}$

C. There are 62 sharks. 45 of the sharks are hungry. How many of the sharks are not hungry?
$\begin{array}{r} {}^{5\,1}\\ \not{6}2 \\ -45 \\ \hline 17 \end{array}$

D. 80 people were watching a bulldozer knock down a building. 69 people left to return to work. How many people stayed to watch?
$\begin{array}{r} {}^{7\,1}\\ \not{8}0 \\ -69 \\ \hline 11 \end{array}$

Part 1 continues on the next page.

Mastery Test Review Lesson 54 (continued)

E. Linda Arrowhead's father is a baker. One Saturday Linda helped her father make pies for some restaurants. They made 60 apple pies and 45 pies that were not apple. How many pies in all did Linda and her father make?
$\begin{array}{r} 60 \\ +45 \\ \hline 105 \end{array}$

2

A. My sister made a beaded belt. She used 140 large beads to make the belt. 90 of the beads are red. How many of the beads are not red?
$\begin{array}{r} 140 \\ -90 \\ \hline 50 \end{array}$

B. There's a new statue in our city. Lots of people saw the statue. 80 people liked the statue. 57 people did not like the statue. How many people saw the statue?
$\begin{array}{r} 80 \\ +57 \\ \hline 137 \end{array}$

C. There are 140 people on the beach. 95 of them are wearing hats. How many of the people are not wearing hats?
$\begin{array}{r} {}^{0\,13\,1}\\ \not{1}\not{4}0 \\ -95 \\ \hline 45 \end{array}$

D. There were 86 earthquakes last year. 29 of the earthquakes were in Europe. How many earthquakes were not in Europe?
$\begin{array}{r} {}^{7\,1}\\ \not{8}6 \\ -29 \\ \hline 57 \end{array}$

E. We made sandwiches for our club's picnic. The people who came ate 95 ham sandwiches and 120 chicken sandwiches. How many sandwiches were eaten at the picnic?
$\begin{array}{r} 95 \\ +120 \\ \hline 215 \end{array}$

Mastery Test Review—Lesson 58

1

A. 11: 4 — 11 − 4 = 7; [7] — 11 − 7 = 4
B. 11: 2 — 11 − 2 = 9; [9] — 11 − 9 = 2
C. 11: 5 — 11 − 5 = 6; [6] — 11 − 6 = 5
D. 11: 3 — 11 − 3 = 8; [8] — 11 − 8 = 3

2

$\begin{array}{r} 11 \\ -8 \\ \hline 3 \end{array}$ $\begin{array}{r} 11 \\ -2 \\ \hline 9 \end{array}$ $\begin{array}{r} 11 \\ -7 \\ \hline 4 \end{array}$ $\begin{array}{r} 11 \\ -4 \\ \hline 7 \end{array}$ $\begin{array}{r} 11 \\ -6 \\ \hline 5 \end{array}$ $\begin{array}{r} 11 \\ -9 \\ \hline 2 \end{array}$ $\begin{array}{r} 11 \\ -3 \\ \hline 8 \end{array}$ $\begin{array}{r} 11 \\ -7 \\ \hline 4 \end{array}$

$\begin{array}{r} 11 \\ -6 \\ \hline 5 \end{array}$ $\begin{array}{r} 11 \\ -4 \\ \hline 7 \end{array}$ $\begin{array}{r} 11 \\ -8 \\ \hline 3 \end{array}$ $\begin{array}{r} 11 \\ -6 \\ \hline 5 \end{array}$ $\begin{array}{r} 11 \\ -5 \\ \hline 6 \end{array}$ $\begin{array}{r} 11 \\ -7 \\ \hline 4 \end{array}$ $\begin{array}{r} 11 \\ -3 \\ \hline 8 \end{array}$ $\begin{array}{r} 11 \\ -8 \\ \hline 3 \end{array}$

Mastery Test Review—Lesson 60

1

A. Mrs. Savas is 42 years old. Miss Hark is 70 years old. How many years older is Miss Hark?
$\begin{array}{r} {}^{6\,1}\\ \not{7}0 \\ -42 \\ \hline 28 \end{array}$

B. After a few games of table tennis, Amos Silverheels had scored 26 points. His brother Mike had scored 40 points. How many points did Amos and Mike score in all?
$\begin{array}{r} 26 \\ +40 \\ \hline 66 \end{array}$

C. Gina weighs 48 kilograms. Sam weighs 29 kilograms. How many more kilograms does Gina weigh?
$\begin{array}{r} {}^{3\,1}\\ \not{4}8 \\ -29 \\ \hline 19 \end{array}$

Part 1 continues on the next page.

Mastery Test Review Lesson 60 (continued)

D. Mel exercised for 48 minutes. Brad exercised for 42 minutes. How many minutes did Mel and Brad exercise in all?
$\begin{array}{r} 48 \\ +42 \\ \hline 90 \end{array}$

E. A building is 36 meters tall. A tree near the building is 48 meters tall. How many meters taller is the tree?
$\begin{array}{r} 48 \\ -36 \\ \hline 12 \end{array}$

2

A. Jana and Allison like to play marbles. Jana has 137 marbles. Allison has 170 marbles. How many more marbles does Allison have than Jana?
$\begin{array}{r} {}^{6\,1}\\ 1\not{7}0 \\ -137 \\ \hline 33 \end{array}$

B. My mother's new car weighs 1340 kilograms. My mother's old car weighed 1700 kilograms. How many kilograms do the cars weigh in all?
$\begin{array}{r} 1340 \\ +1700 \\ \hline 3040 \end{array}$

C. Mr. Mills is 45 years old. Mr. Young is 60 years old. How many years older is Mr. Young?
$\begin{array}{r} {}^{5\,1}\\ \not{6}0 \\ -45 \\ \hline 15 \end{array}$

D. Rafael and Matt worked as ticket sellers at a circus. Rafael sold 94 tickets and Matt sold 120 tickets. How many tickets did the boys sell?
$\begin{array}{r} 94 \\ +120 \\ \hline 214 \end{array}$

E. An apple tree is 594 centimeters tall. An oak tree is 730 centimeters tall. How many centimeters taller than the oak tree is the apple tree?
$\begin{array}{r} {}^{6\,12\,1}\\ \not{7}\not{3}0 \\ -594 \\ \hline 136 \end{array}$

Mastery Test Review—Lesson 63

1

A. 17 { 9: 17 − 9 = 8; [8]: 17 − 8 = 9

B. 14 { 9: 14 − 9 = 5; [5]: 14 − 5 = 9

C. 15 { 9: 15 − 9 = 6; [6]: 15 − 6 = 9

D. 16 { 9: 16 − 9 = 7; [7]: 16 − 7 = 9

2

15 − 9 = 6	17 − 8 = 9	12 − 6 = 6	10 − 5 = 5	16 − 7 = 9	14 − 7 = 7	14 − 9 = 5	15 − 6 = 9
15 − 7 = 8	14 − 5 = 9	14 − 6 = 8	14 − 7 = 7	17 − 8 = 9	17 − 9 = 8	16 − 7 = 9	14 − 5 = 9

Mastery Test Review—Lesson 64

1

A There are 170 birds in a zoo. There are 200 fish in the zoo. How many fewer birds than fish are in the zoo?

[30] birds — $200 - 170 = 30$

B We saw lots of squirrels and rabbits. We counted 143 squirrels and 180 rabbits. How many animals did we see?

[323] animals — $143 + 180 = 323$

Part 1 continues on the next page.

Mastery Test Review Lesson 64 (continued)

C The cook let us choose cake or ice cream for dessert. 145 kids ate cake and 200 kids ate ice cream. How many more kids ate ice cream than cake?

[55] kids — $200 - 145 = 55$

D Iris taught 36 children how to jump rope. Christina taught 40 children how to jump rope. How many children learned how to jump rope?

[76] children — $36 + 40 = 76$

E Chris has 47 rocks. Judd has 82 rocks. How many more rocks does Judd have than Chris?

[35] rocks — $82 - 47 = 35$

F There are 90 bottles of soda pop. 72 of the bottles are full. How many bottles are not full?

[18] bottles — $90 - 72 = 18$

G 1350 railway cars are full. 1500 railway cars are empty. How many railway cars are there in all?

[2850] railway cars — $1350 + 1500 = 2850$

H The Fireside Bookshop sold 1826 books in December. The shop sold 1900 books in January. How many books did the shop sell?

[3726] books — $1826 + 1900 = 3726$

I A hockey team lost 42 games. It played 70 games in all. How many games did the hockey team win?

[28] games — $70 - 42 = 28$

Part 1 continues on the next page.

Mastery Test Review Lesson 64 (continued)

J Fran ran 37 kilometers this week. Carol ran 40 kilometers this week. How many kilometers did Fran and Carol run this week?

[77] kilometers — $37 + 40 = 77$

K Louise made 49 small cakes with strawberry frosting. Then she made some cakes with chocolate frosting. She baked 72 cakes in all. How many cakes had chocolate frosting?

[23] cakes — $72 - 49 = 23$

L Ann read 167 pages of a book. On her day off she read some more. Now Ann has read 210 pages. How many pages did Ann read on her day off?

[43] pages — $210 - 167 = 43$

2

A 64 people walked around the ladder. 140 people walked under the ladder. How many more people walked under the ladder than around the ladder?

[76] people — $140 - 64 = 76$

B In one week 148 new cars and 473 old cars were washed at the Happy Car Wash. How many cars in all were washed?

[621] cars — $148 + 473 = 621$

C At a camp 3802 oranges were eaten in one week. 1648 apples were also eaten. How many more oranges than apples were eaten?

[2154] oranges — $3802 - 1648 = 2154$

Part 2 continues on the next page.

Mastery Test Review Lesson 64 (continued)

D Andy put 2509 boxes on trucks. Randy put 3108 boxes on trucks. How many fewer boxes did Andy put on trucks than Randy?

[599] boxes — $3108 - 2509 = 599$

E Workers chopped down 185 oak trees and 190 pine trees. How many trees were chopped down?

[375] trees — $185 + 190 = 375$

F Last spring a shop had a sale on kites. The shop sold 158 red kites and 190 yellow kites. How many kites did the shop sell?

[348] kites — $158 + 190 = 348$

G Lauren planted 97 flowers. Then she planted some more flowers. Now 102 flowers are planted. How many more flowers did she plant?

[5] flowers — $102 - 97 = 5$

H A newspaper stand received 1444 magazines. 98 of the magazines had torn pages. How many magazines did not have torn pages?

[1346] magazines — $1444 - 98 = 1346$

148 people were ice-skating. Later, more people came to skate. Now there are 190 skaters. How many people came late to skate?

[42] people — $190 - 148 = 42$

Part 2 continues on the next page.

J Phil danced for 148 minutes. Then he danced for another 52 minutes. How many minutes did he dance in all?

$$\begin{array}{r} 148 \\ +\ 52 \\ \hline 200 \end{array}$$

200 minutes

K Workers cleaned 18 airplanes. They cleaned 24 boats. How many things did the workers clean?

$$\begin{array}{r} 18 \\ +\ 24 \\ \hline 42 \end{array}$$

42 things

L An office had 1405 packages of paper. The office used up 1289 packages. How many packages of paper were left?

$$\begin{array}{r} 1405 \\ -\ 1289 \\ \hline 116 \end{array}$$

116 packages of paper

Lesson 3

Name ____________________

6 − 3 = 3	1 − 1 = 0	7 − 1 = 6	2 − 1 = 1	6 − 3 = 3	9 − 1 = 8	3 − 1 = 2
9 − 1 = 8	6 − 4 = 2	3 − 1 = 2	4 − 1 = 3	8 − 1 = 7	5 − 1 = 4	6 − 3 = 3
5 − 1 = 4	6 − 1 = 5	2 − 1 = 1	7 − 1 = 6	6 − 4 = 2	4 − 1 = 3	6 − 4 = 2
6 − 3 = 3	7 − 1 = 6	5 − 1 = 4	6 − 4 = 2	4 − 1 = 3	5 − 1 = 4	8 − 1 = 7
9 − 1 = 8	2 − 1 = 1	3 − 1 = 2	6 − 3 = 3	1 − 1 = 0	8 − 1 = 7	6 − 1 = 5

Lesson 6

Name ____________________

1 − 1 = 0	10 − 5 = 5	7 − 1 = 6	2 − 1 = 1	6 − 3 = 3	18 − 9 = 9	3 − 1 = 2
9 − 1 = 8	6 − 4 = 2	3 − 1 = 2	8 − 4 = 4	18 − 9 = 9	1 − 1 = 0	10 − 5 = 5
5 − 1 = 4	6 − 1 = 5	2 − 1 = 1	10 − 5 = 5	8 − 4 = 4	4 − 1 = 3	6 − 4 = 2
6 − 3 = 3	7 − 1 = 6	18 − 9 = 9	6 − 4 = 2	8 − 4 = 4	10 − 5 = 5	8 − 1 = 7
9 − 1 = 8	6 − 3 = 3	6 − 1 = 5	18 − 9 = 9	8 − 4 = 4	8 − 1 = 7	4 − 1 = 3

Lesson 8

Name ____________________

1 − 0 = 1	6 − 0 = 6	7 − 1 = 6	3 − 0 = 3	6 − 3 = 3	5 − 0 = 5	3 − 1 = 2
9 − 1 = 8	6 − 4 = 2	14 − 7 = 7	4 − 0 = 4	18 − 9 = 9	1 − 0 = 1	10 − 1 = 9
5 − 1 = 4	6 − 1 = 5	2 − 1 = 1	2 − 0 = 2	8 − 4 = 4	4 − 1 = 3	6 − 4 = 2
1 − 1 = 0	4 − 0 = 4	6 − 0 = 6	14 − 7 = 7	5 − 0 = 5	10 − 5 = 5	8 − 1 = 7
8 − 4 = 4	10 − 1 = 9	14 − 7 = 7	18 − 9 = 9	2 − 0 = 2	10 − 5 = 5	3 − 0 = 3

Lesson 10 Name ____________________

6	7	8	10	12	18	15
− 6	− 0	− 8	−10	− 6	− 9	−10
0	7	0	0	6	9	5
8	9	9	12	2	11	16
− 4	− 0	− 9	−10	− 0	−10	− 8
4	9	0	2	2	1	8
4	10	7	13	12	11	8
− 0	−10	− 0	−10	− 6	−10	− 0
4	0	7	3	6	1	8
8	14	10	15	14	16	12
− 8	−10	− 0	−10	− 7	− 8	−10
0	4	10	5	7	8	2
9	13	6	9	14	10	8
− 9	−10	− 6	− 0	−10	− 0	− 0
0	3	0	9	4	10	8

Lesson 13 Name ____________________

6	18	14	14	19	5	9
− 3	− 9	− 7	−10	−10	− 5	− 8
3	9	7	4	9	0	1
2	16	3	6	8	3	19
− 2	−10	− 3	− 5	− 7	− 2	−10
0	6	0	1	1	1	9
5	4	18	7	12	9	6
− 4	− 4	−10	− 7	− 6	− 8	− 5
1	0	8	0	6	1	1
7	4	5	17	10	8	4
− 6	− 3	− 5	−10	− 9	− 7	− 4
1	1	0	7	1	1	0
4	18	3	5	7	7	10
− 3	−10	− 3	− 4	− 6	− 7	− 9
1	8	0	1	1	0	1

Lesson 16 Name ____________________

3	12	16	2	3	14	12
− 1	− 6	− 8	− 2	− 3	− 6	− 4
2	6	8	0	0	8	8
6	10	8	14	12	3	6
− 4	− 5	− 4	− 8	− 2	− 2	− 3
2	5	4	6	10	1	3
4	9	12	8	9	12	18
− 4	− 8	− 3	− 8	− 9	− 5	− 9
0	1	9	0	0	7	9
6	12	14	4	7	12	14
− 5	− 3	− 8	− 3	− 7	− 2	− 6
1	9	6	1	0	10	8
5	12	8	7	11	12	18
− 4	− 4	− 7	− 6	−10	− 5	− 9
1	8	1	1	1	7	9

Lesson 18 Name ____________________

5	14	18	7	14	7	13
− 1	− 7	− 8	− 3	− 4	− 2	− 2
4	7	10	4	10	5	11
3	9	14	12	13	7	5
− 3	− 8	− 6	− 3	− 3	− 2	− 4
0	1	8	9	10	5	1
4	6	14	7	12	14	17
− 3	− 5	− 8	− 7	− 2	− 4	− 7
1	1	6	0	10	10	10
7	8	12	18	7	19	13
− 6	− 7	− 4	− 8	− 3	− 9	− 3
1	1	8	10	4	10	10
2	12	17	19	3	4	5
− 2	− 5	− 7	− 9	− 2	− 4	− 5
0	7	10	10	1	0	0

Lesson 20

Name ______________________

6	4	10	12	12	12	7
− 4	− 1	− 5	−10	− 7	− 9	− 6
2	**3**	**5**	**2**	**5**	**3**	**1**
6	12	17	7	14	12	12
− 3	− 8	−10	− 7	− 8	− 2	− 4
3	**4**	**7**	**0**	**6**	**10**	**8**
8	12	7	12	18	14	17
− 4	− 9	− 3	− 6	− 8	− 4	− 7
4	**3**	**4**	**6**	**10**	**10**	**10**
6	16	12	19	12	12	13
− 6	− 8	− 3	− 9	− 7	− 8	− 3
0	**8**	**9**	**10**	**5**	**4**	**10**
4	12	12	7	12	12	18
− 3	− 5	− 7	− 2	− 9	− 8	− 9
1	**7**	**5**	**5**	**3**	**4**	**9**

Lesson 23

Name ______________________

9	10	7	10	7	12	12
− 9	− 3	− 5	− 4	− 3	− 3	− 9
0	**7**	**2**	**6**	**4**	**9**	**3**
3	5	12	10	14	13	7
− 2	− 4	− 5	− 2	− 7	−10	− 4
1	**1**	**7**	**8**	**7**	**3**	**3**
5	9	14	18	12	14	7
− 5	− 8	− 6	− 8	− 7	− 4	− 5
0	**1**	**8**	**10**	**5**	**10**	**2**
8	6	12	12	10	7	14
− 7	− 5	− 2	− 8	− 4	− 4	− 8
1	**1**	**10**	**4**	**6**	**3**	**6**
7	12	13	10	17	19	10
− 2	− 4	− 3	− 3	− 7	− 9	− 2
5	**8**	**10**	**7**	**10**	**10**	**8**

Lesson 26

Name ______________________

2	12	16	4	7	12	12
− 1	− 6	− 8	− 3	− 7	− 8	− 9
1	**6**	**8**	**1**	**0**	**4**	**3**
6	19	2	10	7	12	7
− 4	−10	− 2	− 2	− 6	− 4	− 2
2	**9**	**0**	**8**	**1**	**8**	**5**
3	10	16	10	12	3	14
− 0	− 9	−10	− 2	− 3	− 3	− 8
3	**1**	**6**	**8**	**9**	**0**	**6**
6	4	10	18	8	15	7
− 6	− 4	− 5	− 9	− 4	−10	− 5
0	**0**	**5**	**9**	**4**	**5**	**2**
8	7	10	14	12	7	12
− 8	− 4	− 4	− 6	− 5	− 3	− 7
0	**3**	**6**	**8**	**7**	**4**	**5**

Lesson 28

Name ______________________

7	11	16	15	12	10	14
− 6	− 9	− 7	− 9	− 9	− 2	− 4
1	**2**	**9**	**6**	**3**	**8**	**10**
7	13	16	12	12	17	12
− 2	− 9	− 9	− 2	− 7	− 9	− 4
5	**4**	**7**	**10**	**5**	**8**	**8**
14	7	14	12	12	10	14
− 6	− 4	− 9	− 9	− 8	− 3	− 9
8	**3**	**5**	**3**	**4**	**7**	**5**
12	17	16	10	12	12	19
− 3	− 9	− 7	− 4	− 9	− 5	− 9
9	**8**	**9**	**6**	**3**	**7**	**10**
11	15	7	12	16	13	14
− 9	− 9	− 5	− 9	− 9	− 9	− 8
2	**6**	**2**	**3**	**7**	**4**	**6**

Lesson 30

Name ______________________

8 − 5 3	12 − 6 6	18 −10 8	5 − 4 1	6 − 5 1	13 − 3 10	17 − 7 10
8 − 3 5	7 − 2 5	7 − 5 2	13 − 9 4	7 − 3 4	8 − 3 5	10 − 4 6
16 − 8 8	8 − 3 5	16 − 9 7	8 − 5 3	11 − 9 2	7 − 4 3	12 − 8 4
18 − 8 10	10 − 2 8	8 − 3 5	16 − 7 9	17 − 9 8	5 − 1 4	14 − 9 5
12 − 7 5	15 − 9 6	8 − 4 4	14 − 7 7	18 − 9 9	12 − 9 3	11 −10 1

Lesson 32

Name ______________________

8 − 5 3	15 − 9 6	7 − 5 2	10 − 3 7	10 − 5 5	6 − 4 2	5 − 5 0
9 − 8 1	12 − 4 8	7 − 2 5	17 − 9 8	16 − 7 9	7 − 4 3	14 − 9 5
5 − 0 5	14 − 6 8	12 − 5 7	7 − 3 4	12 − 9 3	10 − 4 6	11 − 9 2
14 − 8 6	8 − 3 5	14 − 4 10	13 − 9 4	8 − 1 7	14 −10 4	3 − 2 1
16 − 9 7	12 − 3 9	12 − 7 5	8 − 7 1	10 − 9 1	16 −10 6	5 − 4 1

Lesson 34

Name ______________________

5 − 3 2	10 − 8 2	8 − 5 3	17 − 9 8	7 − 4 3	10 − 3 7	12 − 8 4
6 − 6 0	5 − 2 3	10 − 6 4	16 − 9 7	8 − 8 0	11 − 9 2	10 − 7 3
13 − 9 4	4 − 0 4	6 − 5 1	18 − 8 10	11 − 9 2	5 − 3 2	16 − 7 9
10 − 6 4	6 − 1 5	12 − 6 6	19 − 9 10	7 − 5 2	10 − 7 3	5 − 2 3
15 − 9 6	10 − 4 6	8 − 3 5	7 − 6 1	14 − 9 5	10 − 8 2	10 − 2 8

Lesson 36

Name ______________________

9 − 4 5	13 − 8 5	5 − 3 2	10 − 7 3	8 − 5 3	17 − 9 8	7 − 5 2
7 − 2 5	12 − 2 10	13 − 5 8	10 − 8 2	16 − 9 7	10 − 3 7	7 − 4 3
13 − 9 4	9 − 4 5	8 − 3 5	15 − 9 6	10 − 6 4	13 − 5 8	7 − 3 4
17 − 7 10	9 − 2 7	16 − 7 9	14 − 9 5	13 − 8 5	9 − 3 6	12 − 9 3
10 − 4 6	9 − 4 5	5 − 2 3	9 − 2 7	11 − 9 2	12 − 7 5	12 − 4 8

Lesson 38

Name ______________________

7 − 4 3	13 − 7 6	5 − 3 2	16 − 9 7	10 − 3 7	17 −10 7	13 − 3 10
7 − 5 2	13 − 7 6	9 − 3 6	12 − 8 4	4 − 3 1	14 − 9 5	8 − 3 5
13 − 5 8	10 − 7 3	17 − 9 8	8 − 5 3	9 − 2 7	8 − 7 1	16 − 8 8
13 − 8 5	10 − 8 2	15 − 9 6	10 − 4 6	13 − 7 6	5 − 2 3	12 − 9 3
10 − 2 8	5 − 4 1	10 − 6 4	11 − 9 2	16 − 7 9	9 − 4 5	13 − 9 4

Lesson 40

Name ______________________

9 − 5 4	13 − 6 7	9 − 7 2	13 − 8 5	5 − 2 3	11 −10 1	10 − 5 5
9 − 4 5	10 − 6 4	13 − 4 9	9 − 6 3	18 − 9 9	9 − 1 8	14 − 6 8
13 − 4 9	9 − 5 4	5 − 3 2	10 − 7 3	9 − 6 3	13 − 6 7	9 − 9 0
18 − 8 10	8 − 4 4	18 −10 8	9 − 5 4	9 − 7 2	13 − 5 8	9 − 3 6
13 − 6 7	9 − 2 7	10 − 8 2	9 − 6 3	13 − 7 6	9 − 7 2	13 − 4 9

Lesson 42

Name ______________________

9 − 4 5	13 − 6 7	10 − 6 4	9 − 5 4	10 − 7 3	13 − 9 4	10 − 3 7
7 − 0 7	12 −10 2	12 − 6 6	13 − 8 5	13 − 7 6	5 − 2 3	9 − 7 2
13 − 4 9	9 − 6 3	13 − 5 8	14 − 9 5	6 − 3 3	14 − 7 7	7 − 3 4
12 − 8 4	9 − 2 7	13 − 6 7	9 − 6 3	5 − 3 2	10 − 8 2	9 − 3 6
12 − 7 5	7 − 2 5	13 − 4 9	9 − 7 2	14 − 4 10	9 − 5 4	7 − 7 0

Lesson 44

Name ______________________

8 − 3 5	11 − 2 9	10 − 7 3	5 − 3 2	4 − 2 2	11 − 9 2	8 − 2 6
5 − 2 3	16 − 9 7	7 − 4 3	6 − 2 4	9 − 7 2	13 − 8 5	6 − 2 4
12 − 9 3	10 − 4 6	11 − 2 9	9 − 5 4	9 − 2 7	10 − 8 2	8 − 5 3
8 − 2 6	15 − 9 6	7 − 5 2	4 − 2 2	9 − 7 2	9 − 4 5	13 − 5 8
17 − 9 8	9 − 4 5	13 − 4 9	13 − 6 7	16 − 7 9	13 − 7 6	10 − 6 4

Lesson 46

Name ____________________

8 − 2 = 6	10 − 6 = 4	15 − 7 = 8	15 − 8 = 7	10 − 7 = 3	18 − 9 = 9	11 − 2 = 9
4 − 2 = 2	15 − 7 = 8	5 − 3 = 2	9 − 6 = 3	15 − 6 = 9	16 − 9 = 7	13 − 4 = 9
13 − 7 = 6	13 − 5 = 8	11 − 2 = 9	6 − 2 = 4	15 − 6 = 9	9 − 3 = 6	15 − 8 = 7
8 − 3 = 5	15 − 7 = 8	6 − 2 = 4	10 − 8 = 2	13 − 8 = 5	9 − 4 = 5	9 − 7 = 2
15 − 8 = 7	5 − 3 = 2	9 − 2 = 7	8 − 2 = 6	4 − 2 = 2	13 − 6 = 7	9 − 5 = 4

Lesson 48

Name ____________________

8 − 4 = 4	10 − 5 = 5	18 − 9 = 9	15 − 6 = 9	8 − 7 = 1	19 − 9 = 10	16 − 8 = 8
7 − 4 = 3	15 − 7 = 8	9 − 6 = 3	13 − 6 = 7	15 − 6 = 9	12 − 9 = 3	16 − 7 = 9
10 − 3 = 7	15 − 8 = 7	10 − 4 = 6	7 − 5 = 2	11 − 9 = 2	8 − 5 = 3	9 − 7 = 2
15 − 7 = 8	9 − 5 = 4	13 − 9 = 4	6 − 2 = 4	17 − 9 = 8	14 − 9 = 5	13 − 4 = 9
11 − 2 = 9	8 − 2 = 6	15 − 9 = 6	15 − 6 = 9	4 − 4 = 0	10 − 2 = 8	16 − 9 = 7

Lesson 50

Name ____________________

6 − 5 = 1	10 − 1 = 9	13 − 10 = 3	12 − 8 = 4	14 − 5 = 9	11 − 3 = 8	16 − 10 = 6
9 − 8 = 1	14 − 6 = 8	12 − 2 = 10	14 − 8 = 6	11 − 4 = 7	7 − 2 = 5	11 − 5 = 6
12 − 2 = 10	7 − 3 = 4	12 − 5 = 7	13 − 7 = 6	4 − 2 = 2	14 − 5 = 9	18 − 8 = 10
11 − 5 = 6	12 − 3 = 9	19 − 9 = 10	11 − 4 = 7	15 − 6 = 9	12 − 4 = 8	13 − 3 = 10
14 − 5 = 9	11 − 3 = 8	12 − 7 = 5	11 − 4 = 7	11 − 5 = 6	10 − 2 = 8	5 − 4 = 1

Lesson 52

Name ____________________

7 − 6 = 1	14 − 4 = 10	16 − 9 = 7	8 − 3 = 5	15 − 6 = 9	11 − 3 = 8	11 − 4 = 7
7 − 7 = 0	14 − 5 = 9	9 − 7 = 2	5 − 2 = 3	9 − 4 = 5	13 − 8 = 5	8 − 2 = 6
17 − 9 = 8	10 − 6 = 4	9 − 2 = 7	15 − 8 = 7	9 − 5 = 4	13 − 4 = 9	10 − 7 = 3
15 − 6 = 9	14 − 5 = 9	11 − 4 = 7	10 − 8 = 2	6 − 2 = 4	15 − 7 = 8	9 − 6 = 3
11 − 3 = 8	5 − 3 = 2	11 − 9 = 2	4 − 2 = 2	13 − 6 = 7	9 − 3 = 6	11 − 5 = 6

Lesson 54 Name ____________________

6 − 4 = 2	14 − 10 = 4	8 − 8 = 0	11 − 3 = 8	15 − 6 = 9	6 − 3 = 3	11 − 6 = 5
6 − 0 = 6	11 − 7 = 4	2 − 2 = 0	10 − 3 = 7	11 − 2 = 9	13 − 7 = 6	17 − 8 = 9
10 − 4 = 6	10 − 6 = 4	11 − 8 = 3	11 − 6 = 5	7 − 4 = 3	16 − 7 = 9	11 − 7 = 4
12 − 9 = 3	11 − 8 = 3	11 − 4 = 7	7 − 5 = 2	3 − 3 = 0	11 − 5 = 6	11 − 8 = 3
14 − 5 = 9	17 − 8 = 9	11 − 6 = 5	17 − 10 = 7	17 − 8 = 9	11 − 7 = 4	8 − 1 = 7

Lesson 56 Name ____________________

8 − 3 = 5	16 − 9 = 7	10 − 6 = 4	11 − 3 = 8	17 − 8 = 9	11 − 7 = 4	17 − 9 = 8
5 − 2 = 3	15 − 8 = 7	11 − 3 = 8	10 − 7 = 3	8 − 2 = 6	9 − 5 = 4	10 − 8 = 2
11 − 8 = 3	5 − 3 = 2	9 − 4 = 5	6 − 2 = 4	15 − 7 = 8	11 − 6 = 5	17 − 8 = 9
13 − 5 = 8	4 − 2 = 2	11 − 4 = 7	14 − 5 = 9	11 − 6 = 5	13 − 8 = 5	9 − 2 = 7
13 − 6 = 7	11 − 7 = 4	9 − 3 = 6	13 − 4 = 9	9 − 6 = 3	11 − 8 = 3	9 − 7 = 2

Lesson 58 Name ____________________

8 − 4 = 4	18 − 9 = 9	14 − 7 = 7	9 − 9 = 0	12 − 6 = 6	16 − 8 = 8	11 − 8 = 3
4 − 3 = 1	10 − 9 = 1	7 − 2 = 5	7 − 3 = 4	19 − 9 = 10	11 − 7 = 4	12 − 7 = 5
15 − 6 = 9	11 − 3 = 8	17 − 8 = 9	14 − 5 = 9	11 − 6 = 5	11 − 2 = 9	13 − 7 = 6
11 − 9 = 2	8 − 5 = 3	10 − 2 = 8	11 − 5 = 6	13 − 9 = 4	5 − 4 = 1	8 − 7 = 1
15 − 9 = 6	14 − 8 = 6	13 − 3 = 10	14 − 9 = 5	12 − 7 = 5	10 − 2 = 8	12 − 8 = 4

Lesson 60 Name ____________________

6 − 4 = 2	8 − 6 = 2	8 − 3 = 5	10 − 6 = 4	11 − 6 = 5	15 − 8 = 7	9 − 6 = 3
9 − 3 = 6	11 − 7 = 4	4 − 2 = 2	13 − 5 = 8	9 − 4 = 5	13 − 4 = 9	10 − 8 = 2
15 − 7 = 8	8 − 2 = 6	9 − 5 = 4	5 − 2 = 3	15 − 6 = 9	11 − 3 = 8	8 − 6 = 2
11 − 4 = 7	6 − 4 = 2	17 − 8 = 9	14 − 5 = 9	5 − 3 = 2	9 − 2 = 7	9 − 7 = 2
11 − 5 = 6	10 − 7 = 3	13 − 6 = 7	6 − 2 = 4	11 − 8 = 3	8 − 5 = 3	13 − 5 = 8

Lesson 62

Name ______________________

8 − 5 = 3	13 − 7 = 6	11 − 2 = 9	12 − 4 = 8	18 − 8 = 10	14 − 4 = 10	12 − 5 = 7
7 − 4 = 3	10 − 3 = 7	7 − 5 = 2	12 − 9 = 3	10 − 4 = 6	8 − 6 = 2	6 − 4 = 2
14 − 6 = 8	3 − 2 = 1	6 − 5 = 1	9 − 8 = 1	17 − 7 = 10	14 − 8 = 6	12 − 3 = 9
11 − 7 = 4	10 − 5 = 5	11 − 10 = 1	10 − 1 = 9	6 − 6 = 0	8 − 8 = 0	16 − 10 = 6
6 − 4 = 2	8 − 6 = 2	17 − 8 = 9	11 − 6 = 5	11 − 8 = 3	3 − 1 = 2	6 − 3 = 3

Lesson 64

Name ______________________

6 − 4 = 2	11 − 3 = 8	12 − 7 = 5	11 − 9 = 2	9 − 4 = 5	13 − 6 = 7	8 − 6 = 2
6 − 2 = 4	12 − 8 = 4	14 − 7 = 7	13 − 9 = 4	9 − 5 = 4	6 − 2 = 4	15 − 8 = 7
11 − 8 = 3	8 − 6 = 2	14 − 5 = 9	15 − 6 = 9	16 − 9 = 7	9 − 6 = 3	8 − 2 = 6
11 − 6 = 5	14 − 9 = 5	15 − 7 = 8	8 − 3 = 5	11 − 4 = 7	17 − 8 = 9	15 − 9 = 6
9 − 7 = 2	15 − 8 = 7	11 − 5 = 6	6 − 4 = 2	17 − 9 = 8	11 − 7 = 4	10 − 2 = 8

Lesson 2

Name ______________________

5 + 5 = 10	9 + 9 = 18	7 + 7 = 14	6 + 6 = 12	8 + 8 = 16	2 + 2 = 4	3 + 3 = 6

A Jason had 4 sharp pencils.
He sharpened 2 more.
How many sharp pencils did he end up with?

4 + 2 = 6 pencils

B Kate had two cupcakes at the class party.
Her friend gave her another cupcake.
How many cupcakes did Kate have?

2 + 1 = 3 cupcakes

C Lisa answered 55 questions on her test.
She took a short break. When she came back, she answered 20 more questions.
How many test questions did Lisa answer?

55 + 20 = 75 questions

450 + 231 = 681	975 + 100 = 1075	642 + 357 = 999	200 + 600 = 800	199 + 800 = 999
(carry 11) 458 + 569 = 1027	(carry 1) 744 + 365 = 1109	(carry 11) 985 + 288 = 1273	(carry 11) 578 + 782 = 1360	(carry 1) 990 + 364 = 1354

Lesson 4

Name ______________________

2 + 2 = 4	3 + 3 = 6	4 + 4 = 8	5 + 5 = 10	6 + 6 = 12	7 + 7 = 14	8 + 8 = 16

A John's dad fixed 3 cars.
He fixed 2 bikes, and then he fixed 8 more cars. How many cars did John's dad fix?

3 + 8 = 11 cars

B Jenny worked 8 hours on Monday.
She worked 12 hours on Tuesday.
She worked 15 hours on Wednesday.
She took the rest of the week off. How many hours did Jenny work in all?

15 + 12 + 8 = 35 hours

C Last year, 4587 lemmings ran to the sea.
This year, 2638 more lemmings than last year are going to run to the sea. How many lemmings are going to run to the sea this year?

(carry 111) 4587 + 2638 = 7225 lemmings

(carry 11) 699 + 587 = 1286	364 + 125 = 489	(carry 11) 457 + 965 = 1422	450 + 200 = 650	(carry 11) 788 + 877 = 1665
(carry 111) 6457 + 9874 = 16331	3612 + 5362 = 8974	(carry 11) 5984 + 8281 = 14265	(carry 1) 3216 + 1478 = 4694	(carry 11) 6940 + 3478 = 10418

Lesson 6

Name ______________________________

3	2	5	1	4	7	6
+ 3	+ 2	+ 5	+ 1	+ 4	+ 7	+ 6
6	**4**	**10**	**2**	**8**	**14**	**12**

1	11	11	11	11
320	436	547	764	357
478	132	640	541	743
+ 241	+ 145	+ 326	+ 225	+ 745
1039	**713**	**1513**	**1530**	**1845**

1	1 1	1 1		
2456	4875	8543	4589	3214
+7641	+6419	+1547	+1210	+3475
10097	**11294**	**10090**	**5799**	**6689**

A Janet had 50 dollars. She earned 75 more raking leaves. How much money did Janet have altogether?

50
+ 75
125 dollars

B Micah sprinted 100 meters in 30 seconds. He jogged for 60 more seconds. Then he walked for 90 seconds. How many seconds did Micah exercise?

30
60
+ 90
180 seconds

C Carlos turned in 2 homework assignments. He finished 4 class projects. He ate lunch. Then he turned in 2 more homework assignments. How many homework assignments did Carlos turn in?

2
+ 2
4 assignments

Lesson 8

Name ______________________________

6	5	8	7	2	4	9
+ 6	+ 5	+ 8	+ 7	+ 2	+ 4	+ 9
12	**10**	**16**	**14**	**4**	**8**	**18**

A Jorge ran 100 yards. Then he ran 25 yards. The quarterback threw him the ball again and he ran another 75 yards. How many yards did Jorge run altogether?

11
100
25
+ 75
200 yards

B Mike was doing dishes after dinner. He washed 4 plates, 3 cups, and 20 pieces of silverware. He wiped down the counter and remembered he had to wash 3 pans. How many dishes in all did Mike end up washing?

1
20
4
3
+ 3
30 dishes

C Alisha had 5 green pencils. She had 2 red pens. She had 7 sheets of paper. She had 9 red pencils. She also had 12 yellow pencils. How many pencils did Alisha have?

1
12
5
+ 9
26 pencils

		11		11
400	450	854	471	688
+ 952	+ 249	+ 148	+ 124	+ 367
1352	**699**	**1002**	**595**	**1055**

111	121	112	111	121
4361	1374	8317	9176	8254
6475	2497	6478	8467	7496
+ 1425	+ 6475	+ 3469	+ 4632	+ 3697
12261	**10346**	**18264**	**22275**	**19447**

Lesson 10

Name ______________________________

10	8	18	5	14	6	6
− 5	− 4	− 4	− 1	− 7	− 3	− 0
5	**4**	**14**	**4**	**7**	**3**	**6**

A Ken had 25 carrots. He ate 12 of them. He ate 2 potatoes. Then he ate 12 more carrots. How many carrots did Ken eat?

12
+12
24 carrots

B Mr. Jackson assigned 3 pages of reading for homework on Monday. On Wednesday he assigned 5 more pages of reading. On Thursday he assigned 9 pages of reading. How many pages of reading did Mr. Jackson assign for homework?

3
5
+ 9
17 pages

C A suit coat cost 450 dollars. A shirt cost 142 dollars. A pair of suit pants cost 120 dollars. If a tie cost 80 dollars, how much did the entire outfit cost?

1
450
142
120
+ 80
792 dollars

11	11	1	11	11
644	249	412	766	245
232	541	314	485	750
+ 128	+ 369	+ 258	+ 844	+ 399
1004	**1159**	**984**	**2095**	**1394**

11	11	11	1	111
6476	2497	6478	8467	7496
+1425	+6475	+3469	+4632	+3697
7901	**8972**	**9947**	**13099**	**11193**

Lesson 12

Name ______________________________

8	8	4	5	7	9	3
− 0	− 1	− 0	− 0	− 1	− 0	− 1
8	**7**	**4**	**5**	**6**	**9**	**2**

1		1	11	11
440	652	896	455	253
+ 270	+ 842	+ 610	+ 196	+ 187
710	**1494**	**1506**	**651**	**440**

1	1	1	11	111
6425	8421	1910	8461	1217
412	750	200	1200	362
+2412	+ 320	+1326	+ 456	+2547
9249	**9491**	**3436**	**10117**	**4126**

A Joanna had 230 motorcycles in her shop. She had 20 cars. She had 410 bicycles. She bought 159 more motorcycles for her shop. How many motorcycles did Joanna have in her shop?

230
+159
389 motorcycles

B Gina was 4 feet tall. She grew another 2 feet in a single year. How tall was Gina at the end of the year?

4
+ 2
6 feet

C Joe was a cook in a café. On Saturday night he made 298 hamburgers. On Sunday night he made 100 more hamburgers than he made on Saturday. How many hamburgers did Joe make on Sunday night?

298
+100
398 hamburgers

Lesson 14

Name ____________________

7 − 1 6	3 − 1 2	9 − 0 9	8 − 1 7	4 − 0 4	7 − 1 6	8 − 0 8

480 − 270 210	774 − 161 613	11 545 + 698 1243	6521 − 1410 5111	1 532 + 874 1406
4412 − 2412 2000	845 − 413 432	1200 − 1100 100	861 − 740 121	9856 − 7535 2321

A Jackie made 35 dollars on Monday. She made 18 dollars on Tuesday. She took Wednesday off, but went back to work on Thursday and made 40 dollars. How much money did Jackie make?

1
35
18
+40
93 dollars

B Alishia wanted to invite 15 friends to her party. Her mother insisted she invite 32 of her cousins. Her father insisted she invite her 20 aunts and uncles. How many people did Alishia invite to her party?

15
32
+20
67 people

C John had 4 pencils, 3 pens, 8 erasers, and 12 highlighters. How many things did John have to write with?

12
4
+ 3
19 things

Lesson 16

Name ____________________

7 − 1 6	7 − 3 4	7 − 2 5	7 − 0 7	7 − 1 6	7 − 2 5	7 − 3 4

A Latisha took 3 tests on Monday. She turned in 5 homework assignments. She took 8 tests on Friday and turned in 9 more homework assignments. How many homework assignments did Latisha turn in?

5
+ 9
14 assignments

B A new computer cost 800 dollars. The software to run the computer cost 450 dollars. A monitor cost 560 dollars. How much did the entire system cost?

1
800
450
+ 560
1810 dollars

C Mike had 75 CDs in his collection. He gave his friend Dave 20 baseball cards. Dave gave Mike 240 CDs. How many CDs did Mike end up with?

1
240
+ 75
315 CDs

467 − 125 342	5413 − 3102 2311	1 413 + 258 671	7415 − 5403 2012	7748 − 3535 4213
985 − 721 264	1 742 + 549 1291	745 − 512 233	6947 − 5835 1112	1 129 + 369 498

Lesson 18

Name ____________________

12 − 6 6	16 − 8 8	14 − 10 4	7 − 1 6	18 − 9 9	8 − 4 4	6 − 3 3

A Nicole had 4 fish in an aquarium. If she added 56 fish, how many fish would she end up with?

1
56
+ 4
60 fish

B Jolene was an ironworker. During week 1 she made 423 welds. During week 2 she made 45 more welds than she had during week 1. How many welds did Jolene make during week 2?

423
+ 45
468 welds

C A pair of socks cost 12 dollars. Shoes cost 50 dollars. A shirt cost 14 dollars. Jeans cost 25 dollars. How much does it cost to buy a pair of socks and shoes?

12
+50
62 dollars

467 − 25 442	8764 − 532 8232	1 524 + 58 582	634 − 23 611	6241 − 5130 1111
981 − 20 961	1 544 + 782 1326	7456 − 200 7256	4375 − 1234 3141	1 540 + 989 1529

Lesson 20

Name ____________________

14 − 10 4	10 − 5 5	6 − 3 3	8 − 4 4	18 − 10 8	16 − 8 8	18 − 9 9

9872 − 251 9621	3535 − 500 3035	111 6452 + 2879 9331	6453 − 322 6131	9869 − 8 9861
1 4526 + 29 4555	1 445 + 25 470	7456 − 6205 1251	4296 − 34 4262	1 450 + 898 1348

A 455 students went to Bright Elementary School. 323 students went to Smart Elementary School. Bright School got 47 new students. 25 students moved away from Smart School. 55 new students started at Bright School. How many students went to Bright Elementary?

11
455
47
+ 55
557 students

B Paulo was a mailman. On Monday he delivered 744 letters. On Tuesday he delivered only 20. On Wednesday he delivered 546 letters. On Thursday he delivered 760 letters. He delivered 87 letters on Friday. How many letters did Paulo deliver that week?

21
744
20
546
760
+ 87
2157 letters

Lesson 22

Name ______________________

9 − 0 = 9	3 − 1 = 2	12 − 6 = 6	18 − 9 = 9	4 − 0 = 4	7 − 1 = 6	6 − 3 = 3

7586 + 145 + 248 = 7979	412 + 764 + 203 = 1379	6487 + 365 + 35 = 6887	7319 + 7216 + 9461 = 23996	5222 + 243 + 754 = 6219
2141 − 21 = 2120	92 − 11 = 81	800 − 500 = 300	645 − 532 = 113	9856 − 35 = 9821

A Tino's best pizza had 744 slices of pepperoni. Their next best pizza had 540 slices of pepperoni. Their worst pizza only had 12 slices of pepperoni. If you ate all three pizzas, how many slices of pepperoni would you have to eat?

744 + 540 + 12 = 1296 slices

B Janelle spent 3 days taking her dad's car apart. She spent 5 days working on the engine. She spent 2 days sleeping. Then she took 3 days to put the car back together. How many days did she work on her dad's car?

3 + 5 + 3 = 11 days

C Mom had 250 dollars. Dad had 30 dollars. David had 575 dollars, but his sister only had 3 dollars. How much money did David's family have altogether?

250 + 30 + 575 + 3 = 858 dollars

Lesson 24

Name ______________________

10 − 5 = 5	8 − 4 = 4	6 − 3 = 3	12 − 6 = 6	14 − 7 = 7	16 − 8 = 8	18 − 9 = 9

7583 − 5367 = 2216	458 − 397 = 61	6565 − 1484 = 5081	860 − 711 = 149	4627 − 1545 = 3082
4321 − 2110 = 2211	452 − 416 = 36	6082 − 1100 = 4982	4147 − 3409 = 738	958 − 555 = 403

A Michael wrote a song that was 15 minutes long. He also wrote songs that were 23, 154, and 98 minutes long. If Michael played all of his songs at a concert, how long would it take?

15 + 23 + 154 + 98 = 290 minutes

B Jake spent 3 hours playing video games. He took 10 minutes to eat lunch. Then he played for 12 more hours. How long did Jake spend playing video games?

12 + 3 = 15 hours

C Gladys' class had 22 students. If 15 more students joined the class, how many students would there be?

22 + 15 = 37 students

Lesson 26

Name ______________________

60 − 1 = 59	30 − 1 = 29	20 − 1 = 19
50 − 1 = 49	90 − 1 = 89	40 − 1 = 39
70 − 1 = 69		

620 − 450 = 170	9852 − 7438 = 2414	8412 − 4351 = 4061	2341 − 1430 = 911	212 − 151 = 61
5694 − 2915 = 2779	8554 − 4736 = 3818	6784 − 5818 = 966	9324 − 1508 = 7816	3895 − 1989 = 1906

A 8 fish were swimming in a fish bowl with 12 snails and 3 frogs. How many creatures were in the fish bowl?

8 + 12 + 3 = 23 creatures

B Tona built a canoe that was 12 feet long. He also built a canoe that was 15 feet long. If the canoes were placed end to end, how long would they be altogether?

12 + 15 = 27 feet

C Samantha wanted to cut her own hair, so she got a pair of scissors and cut off 3 inches. She wasn't happy with the haircut, so she cut off another 5 inches. After looking in the mirror, she cut off 1 more inch of hair. How many inches of hair did she cut off altogether?

3 + 5 + 1 = 9 inches

Lesson 28

Name ______________________

12 − 6 = 6	12 − 4 = 8	12 − 7 = 5	12 − 5 = 7	12 − 3 = 9	12 − 9 = 3	12 − 10 = 2

A Dr. Ramsey gave 3 shots on Monday. She gave 8 shots on Tuesday. She played 18 holes of golf on Wednesday and Thursday. On Friday she gave 2 shots. How many shots did Dr. Ramsey give during the week?

3 + 8 + 2 = 13 shots

B Robin wrote 20 words in 3 minutes during his first writing test. On his second writing test, he wrote 35 words in 3 minutes. How many words did Robin write in all?

20 + 35 = 55 words

C Justin bought 6 guitar strings from the music store. He broke 4 banjo strings. His friend gave him 12 guitar strings. How many guitar strings did Justin get?

6 + 12 = 18 strings

746 − 329 = 417	9463 − 7735 = 1728	625 + 558 = 1183	634 − 523 = 111	6844 − 5130 = 1714
450 − 122 = 328	514 + 762 = 1276	9854 − 7613 = 2241	8544 − 7835 = 709	325 + 898 = 1223

Lesson 30

Name ____________________

12 − 9 = 3   14 − 7 = 7   12 − 7 = 5   16 − 8 = 8   12 − 10 = 2   18 − 9 = 9   10 − 5 = 5

A New Movie Studios made 4 scary movies during their first year. In their second year they made 25 scary movies. They made 60 scary movies during their third year. Then they went out of business. How many scary movies did New Movie Studio make?

4 + 60 + 25 = 89 movies

B On a nature hike, Amanda saw 62 bugs. She saw 3 deer. She hiked 4 miles. She saw 28 more bugs. She saw 1 very large bear. How many bugs did Amanda see on her nature hike?

62 + 28 = 90 bugs

C Danae answered 20 questions on her math test. She took a break and ate 3 peanuts. She came back and answered 10 more questions. Then she ate 9 more peanuts. How many test questions did Danae answer?

20 + 10 = 30 questions

852 − 281 = 571   6482 + 3642 = 10124   265 + 143 = 408   4381 − 1534 = 2827   4846 − 3190 = 1656

540 − 250 = 290   4516 + 3002 = 7518   8594 − 4797 = 3797   5764 − 3794 = 1970   625 − 552 = 73

Lesson 32

Name ____________________

12 − 10 = 2   18 − 9 = 9   16 − 8 = 8   5 − 0 = 5   8 − 1 = 7   12 − 8 = 4   8 − 4 = 4

9874 − 2915 = 6959   3644 + 3644 = 7288   2475 − 146 = 2329   85 − 16 = 69   9486 − 6557 = 2929

504 + 95 = 599   3491 + 5203 = 8694   9978 − 59 = 9919   4661 − 2752 = 1909   90 − 52 = 38

[7] {4, 3}   4 + 3 = 7

9 {2, [7]}   9 − 2 = 7

14 {8, [6]}   14 − 8 = 6

[9] {8, 1}   8 + 1 = 9

Lesson 34

Name ____________________

10 − 5 = 5   18 − 10 = 8   16 − 8 = 8   8 − 4 = 4   12 − 10 = 2   7 − 1 = 6   14 − 7 = 7

A Gerardo made 2706 toothpicks from 1 very small tree. He made 8654 toothpicks from 1 medium sized tree. He made another 2751 toothpicks from another small tree. How many toothpicks did Gerardo make?

2706 + 8654 + 2751 = 14111 toothpicks

B Joanie was digging worms to go fishing. She found 8 worms in the first hole she dug. In the second hole she found 5 beetles. In the third hole she got 10 worms. She also found 2 rocks. How many worms did Joanie find?

8 + 10 = 18 worms

C Vic sold used vehicles. He sold a car for 5500 dollars. He sold a motorcycle for 400 dollars. He sold a truck for 2300 dollars. How much money did Vic make?

5500 + 400 + 2300 = 8200 dollars

912 − 851 = 61   5487 + 3575 = 9062   2600 − 1903 = 697   4736 − 3829 = 907   454 + 231 = 685

9864 − 250 = 9614   2258 − 1749 = 509   785 − 69 = 716   3333 − 3331 = 2   562 + 748 = 1310

Lesson 36

Name ____________________

20 − 1 = 19   80 − 1 = 79   70 − 1 = 69

60 − 1 = 59   30 − 1 = 29   10 − 1 = 9

50 − 1 = 49

[11] {5, 6}   5 + 6 = 11

9 {4, [5]}   9 − 4 = 5

A On a scavenger hunt, Niko found 23 blue bottles. He found 9 telephones. He found 19 green bottles. He found 5 lamp shades. He found 72 clear bottles. How many bottles did Niko find?

23 + 19 + 72 = 114 bottles

780 − 692 = 88   7169 − 6588 = 581   8761 − 6596 = 2165   4621 − 3750 = 871   868 − 199 = 669

6354 − 296 = 6058   7685 − 4894 = 2791   462 + 78 = 540   9340 − 7279 = 2061   645 + 644 = 1289

Lesson 38

Name ____________

15 − 9 = 6	12 − 5 = 7	11 − 9 = 2	14 − 9 = 5	10 − 6 = 4	7 − 5 = 2	18 − 9 = 9

A Randy bought 25 apples at the store. He ate 20 of them. How many apples did he have left?

25 − 20 = 5 apples

B Rachele did 2 jumping jacks, 9 push-ups, and 25 sit-ups. How many exercises did Rachele do?

25 + 9 + 2 = 36 exercises

C Ramon took 45 pictures. He developed 25 of them. How many pictures did Ramon still need to develop?

45 − 25 = 20 pictures

645 − 258 = 387	9745 − 79 = 9666	5743 − 2806 = 2937	65 − 38 = 27	376 + 402 = 778
7831 − 9 = 7822	3794 − 2905 = 889	1946 + 855 = 2801	8437 − 3572 = 4865	229 + 374 = 603

Lesson 40

Name ____________

16 − 9 = 7	11 − 9 = 2	14 − 9 = 5	13 − 9 = 4	17 − 9 = 8	15 − 9 = 6	18 − 9 = 9

A Kresta had 9 cars. She sold 2 of them. How many cars did she have left?

9 − 2 = 7 cars

B Mike was a jockey. He rode 3 horses on Monday. On Tuesday he rode 9 horses. He rode 2 more horses on Wednesday. How many horses did Mike ride?

3 + 9 + 2 = 14 horses

C Shaniqua had 20 dollars. She bought a book for 13 dollars. How many dollars did she have left?

20 − 13 = 7 dollars

1975 − 863 = 1112	6734 − 5963 = 771	280 − 199 = 81	42 + 55 = 97	5487 − 99 = 5388
3871 + 4129 = 8000	3265 − 1198 = 2067	990 + 487 = 1477	8317 − 2576 = 5741	5462 − 395 = 5067

Lesson 42

Name ____________

10 − 6 = 4	10 − 9 = 1	10 − 7 = 3	10 − 3 = 7	10 − 8 = 2	12 − 5 = 7	12 − 3 = 9

A Jason had 2 screwdrivers, 3 hammers, and 2 saws. He had 1 car. He also had 5 pairs of shoes. He had 8 wrenches. How many tools did Jason have?

2 + 3 + 2 + 8 = 15 tools

B Michelle went to the store and bought 12 oranges. She ate 3 of them. How many oranges did she have left?

12 − 3 = 9 oranges

C Phillipe put 29 gallons of gas into his truck. He drove 450 miles and used 28 gallons. How many gallons of gas were left?

29 − 28 = 1 gallon

1678 − 981 = 697	7315 + 98 = 7413	6719 − 1990 = 4729	4255 − 3199 = 1056	8430 − 7310 = 1120
2525 + 4672 = 7197	8517 − 6751 = 1766	9901 − 5893 = 4008	3231 − 1980 = 1251	87 − 59 = 28

Lesson 44

Name ____________

13 − 9 = 4	13 − 5 = 8	13 − 7 = 6	13 − 4 = 9	13 − 8 = 5	13 − 6 = 7	13 − 9 = 4

4550 − 3870 = 680	1546 − 1538 = 8	2830 + 1800 = 4630	8744 − 7944 = 800	9431 + 8 = 9439
1726 + 4400 = 6126	1986 − 542 = 1444	4443 − 3335 = 1108	7453 − 1793 = 5660	6541 − 5900 = 641

A 157 children took a math test. 150 of the children passed the test, but the rest didn't. How many children didn't pass the test?

157 − 150 = 7 children

B Nathan was digging holes in his sandbox. He dug 48 holes. He filled 18 of the holes. How many holes were left?

48 − 18 = 30 holes

C Mr. Macintosh had saxophones and bagpipes. He had 25 instruments. 12 of the instruments were saxophones. How many of the instruments were bagpipes?

25 − 12 = 13 bagpipes

Lesson 46

Name ____________

13 − 6 = 7; 12 − 9 = 3; 13 − 7 = 6; 15 − 9 = 6; 13 − 4 = 9; 17 − 9 = 8; 13 − 5 = 8

A Carlos had 70 gel pens. He used the ink in 25 gel pens drawing a giant picture. How many gel pens did he have left?

70 − 25 = 45 gel pens

B Nina loved fruit. She kept apples and bananas in her refrigerator. Last week Nina had 342 pieces of fruit in her refrigerator. 139 of them were bananas. How many apples did Nina have?

342 − 139 = 203 apples

C Johann ran marathons. Last summer he ran three races. His first race was 20 miles. His second race was 24 miles. His third race was 30 miles. How many miles did Johann run in races last summer?

20 + 24 + 30 = 74 miles

3274 − 1683 = 1591; 6455 + 745 = 7200; 8844 − 7951 = 893; 25 − 19 = 6; 9375 − 7793 = 1582

1452 − 31 = 1421; 450 − 179 = 271; 6987 − 4000 = 2987; 200 − 109 = 91; 7452 − 98 = 7354

Lesson 48

Name ____________

9 − 4 = 5; 9 − 3 = 6; 9 − 6 = 3; 9 − 5 = 4; 9 − 3 = 6; 9 − 6 = 3; 9 − 4 = 5

8543 − 3265 = 5278; 7588 − 1200 = 6388; 3369 − 1881 = 1488; 699 − 511 = 188; 9907 − 9372 = 535

1875 + 6845 = 8720; 324 − 84 = 240; 2831 − 1951 = 880; 7634 − 6523 = 1111; 212 + 121 = 333

A Mitch was a waiter at the Hungry Duck restaurant. He made 45 dollars in tips on Saturday. He made 22 dollars during the morning. How many dollars did he make in the evening?

45 − 22 = 23 dollars

B Chauncy, the dog, did tricks. Chauncy rolled over 3 times, sat up 9 times, and shook hands with 325 people. How many tricks did Chauncy do?

3 + 9 + 325 = 337 tricks

C Joe brought 190 sharp pencils to school for the big writing test. If he broke 87 of them during the first 10 minutes of the test, how many pencils would he have left?

190 − 87 = 103 pencils

Lesson 50

Name ____________

9 − 6 = 3; 10 − 3 = 7; 9 − 4 = 5; 10 − 6 = 4; 9 − 5 = 4; 10 − 7 = 3; 9 − 3 = 6

255 − 197 = 58; 8425 − 1754 = 6671; 2741 − 1178 = 1563; 898 − 99 = 799; 3674 − 1286 = 2388

5642 + 478 + 1975 = 8095; 3600 + 2411 + 9741 = 15752; 47 + 64 + 36 = 147; 8743 + 6412 + 320 = 15475; 450 + 9 + 187 = 646

A. 3256 people waited in line for tickets to the football game. More people showed up later. When the ticket counter opened, 4000 people were in line. How many people showed up later?

4000 − 3256 = 744 people

B. Mrs. Fancy had 24 dress coats in her closet. Her cat tore 14 of them. How many coats did she have left?

24 − 14 = 10 coats

C. Randy collected baseball cards. He had 298 cards in his collection. After trading with Jennifer, he had 487 cards in his collection. How many cards did Randy get from Jennifer?

487 − 298 = 189 cards

Lesson 52

Name ____________

15 − 6 = 9; 13 − 6 = 7; 15 − 7 = 8; 13 − 4 = 9; 15 − 8 = 7; 13 − 8 = 5; 15 − 9 = 6

A Savannah had 12 sandwiches for lunch. 9 of them were peanut butter. The rest were anchovy. How many anchovy sandwiches did Savannah have?

12 − 9 = 3 anchovy sandwiches

B Cody picked up 239 cans from the side of the road on Saturday. On Sunday he picked up more cans. He ended up with 562 cans. How many cans did Cody pick up on Sunday?

562 − 239 = 323 cans

C Mikayla worked 12 hours on Monday, 10 hours on Tuesday, and 8 hours on Wednesday. She went to the store 12 times on Thursday. She worked 8 hours on Friday. How many hours did Mikayla work?

12 + 10 + 8 + 8 = 38 hours

8135 − 7584 = 551; 2341 + 1654 = 3995; 3640 − 2510 = 1130; 2539 − 1990 = 549; 7234 − 7134 = 100

6721 − 5653 = 1068; 4592 − 1793 = 2799; 1544 + 6877 = 8421; 2302 − 1098 = 1204; 7452 + 598 = 8050

Lesson 54 Name ____________________

13	16	4	15	5	11	9
− 7	− 8	− 2	− 7	− 4	− 2	− 2
6	8	2	8	1	9	7

699	099	199	799	299
7000	1000	2000	8000	3000
−1236	− 934	−1465	−6477	− 5
5764	66	535	1523	2995

511	1 1		5 12	2 10
621	9716	1700	6372	6311
− 154	+ 616	−1200	− 991	−1247
467	10332	500	5381	5064

A Stuart was 12 years old. He got older. He is now 24 years old. How many years older is Stuart?

24 − 12 = 12 years

B There were 32 students in a classroom. 15 students were boys. How many students in the classroom were girls?

(2) 32 − 15 = 17 girls

C Jorge had 25 suits cleaned. The cleaners lost some of his suits. He picked up 17 suits from the cleaners. How many suits did the cleaners lose?

(1) 25 − 17 = 8 suits

Lesson 56 Name ____________________

10	6	15	10	9	15	7
− 4	− 2	− 7	− 3	− 2	− 8	− 2
6	4	8	7	7	7	5

A Mr. Mendez had 35 items in his cart. 18 of the items were boxes. How many items were not boxes?

(2) 35 − 18 = 17 items

B Francine had 20 dresses. She went shopping and bought more dresses. She ended up with 32 dresses. How many dresses did Francine buy?

32 − 20 = 12 dresses

C Patrick had 725 video games. 430 of the games were driving games. How many of the video games were not driving games?

(6) 725 − 430 = 295 games

199		211	799	19
2000	3415	3254	8004	6200
− 765	+2254	−1981	−7545	− 34
1235	5669	1273	459	6166

399	1 1	11	216	89
4006	2738	9754	3724	9007
−3958	+6452	+ 57	−1954	− 654
48	9190	9811	1770	8353

Lesson 58 Name ____________________

17	16	12	14	14	13	14
− 8	− 7	− 9	− 5	− 9	− 4	− 5
9	9	3	9	5	9	9

A Gordon ordered 243 boxes of lettuce. He ordered 420 boxes of carrots. He ordered 19 boxes of onions. How many boxes of vegetables did Gordon order?

(1) 243 + 420 + 19 = 682 boxes

B There are 20 balls in Daishon's garage. 12 of the balls are footballs. How many of the balls are not footballs?

(1) 20 − 12 = 8 balls

C The Shipmates were a rock group. 2578 people showed up to their concert early. The rest of the people showed up late. 8120 people were at the concert. How many people showed up late?

(7 10 11) 8120 − 2578 = 5542 people

6 9 13	8	0 9 9	1	4 9 9
7040	1908	1000	6540	5000
−1562	−1642	− 999	+1374	−1886
5478	266	1	7914	3114

	1	211		811
6100	3791	325	8000	9237
+2574	+2808	− 57	+7641	−3651
8674	6599	268	15641	5586

Lesson 60 Name ____________________

10	7	11	14	15	12	18
− 6	− 5	− 9	− 9	− 9	− 5	− 9
4	2	2	5	6	7	9

A During her vacation, Rowena drove 700 miles to get to the amusement park. She drove 987 miles to get to the beach. How many miles did Rowena drive?

700 + 987 = 1687 miles

B A car dealer had 54 vehicles. 22 of the vehicles were trucks. How many of the vehicles were not trucks?

54 − 22 = 32 vehicles

C Candace was 4 feet tall when she was 5 years old. She grew taller. When she was 9 years old, she was 5 feet tall. How many feet did Candace grow?

5 − 4 = 1 foot

5 9 9	8 9	51310	6 13	1 1
6000	2904	6412	3741	6204
−4571	−1849	−5535	−1374	+1988
1429	1055	877	2367	8192

	71010	1 6	8119	21113
560	8112	2572	9200	3243
+ 224	−3658	−1947	−7621	− 569
784	4454	625	1579	2674

Lesson 62 Name ______________________________

16	7	18	12	14	6	8
− 8	− 1	− 9	− 6	− 10	− 3	− 4
8	**6**	**9**	**6**	**4**	**3**	**4**

9565	1700	4212	9764	3358
− 2800	− 1655	− 88	+ 520	+ 7841
6765	**45**	**4124**	**10284**	**11199**

6102	9000	2010	6852	5800
− 2348	− 5832	− 1275	+ 9871	− 374
3754	**3168**	**735**	**16723**	**5426**

A Gretchen ordered 25 CDs from the CD of the month club. Her sister ordered 52 CDs. How many more CDs did Gretchen's sister order?

52 − 25 = **27 CDs**

B If there were 4897 people at a party and 2471 left, how many would still be there?

4897 − 2471 = **2426 people**

C Jimmy Anderson was a baseball star. He hit 200 home runs during his first year. He hit 412 home runs during his second year. How many more home runs did Jimmy Anderson hit during his second year?

412 − 200 = **212 home runs**

Lesson 64 Name ______________________________

14	14	14	17	17	12	16
− 5	− 7	− 9	− 8	− 9	− 6	− 7
9	**7**	**5**	**9**	**8**	**6**	**9**

6874	2974	6900	2714	3641
− 5500	− 1891	− 4525	− 1736	− 2989
1374	**1083**	**2375**	**978**	**652**

3255	6546	4724	8246	874
1984	1873	87	3497	41
+ 5791	+ 1973	+ 3624	+ 730	+ 137
11030	**10392**	**8435**	**12473**	**1052**

A There were 1647 bottles of cranberry juice at the store. There were also 1974 bottles of apple juice. How many more bottles of apple juice were there?

1974 − 1647 = **327 bottles**

B Randall put together 5246 model cars. He put together more model cars. After he was finished, he had 9734 model cars. How many new model cars did Randall put together?

9734 − 5246 = **4488 model cars**

C Jolene was watching television. On Monday night she watched television for 240 minutes. There were 180 minutes of commercials. How many minutes were not commercials?

240 − 180 = **60 minutes**